Ed Fowler's
Knife Talk II
The High Performance Blade

© 2003
by Ed Fowler
Printed in the United States of America.
All rights reserved.

No portion of this publication may be reproduced or transmitted in any form or by any means, electronic or mechanical, including photocopy, recording, or any information storage and retrieval system, without permission in writing from the publisher, except by a reviewer who may quote brief passages in a critical article or review to be printed in a magazine or newspaper, or electronically transmitted on radio or television.

Published by

700 East State Street • Iola, WI 54990-0001
715-445-2214 • 888-457-2873
www.krause.com

Please call or write for our free catalog of publications. Our toll-free number to place an order or obtain a free catalog is 800-258-0929 or please use our regular business telephone, 715-445-2214.

Library of Congress Catalog Number: 97-80612

ISBN: 0-87349-564-0

Edited by: Dennis Thornton
Designed by: Ethel Thulien

Acknowledgments

First of all, thanks to Angie. Without her dedication to making them happen, even the first book would never have been put together. Also to Dave, Rex, Steve, Bill, Tracy, and the many knife makers from times past who have shared thoughts with me through their knives; HDT, who shared his thoughts in his journals and proved to me that I am not crazy for the thoughts I have in seeking to know the world around me.

This book would never have come to be without the contribution and welcomed active support of you, the readers of *Blade* and *Knife Talk*, and those who have contacted me to share time about knives. No matter how simple or complex the knife, it always shares time with those very special creatures, human beings, who occupy the exact center of the world of knives. Together, we have built a community that is a nice place to be.

Steve Shackleford, editor of *Blade*, and I continually lock horns as these articles develop from my thoughts about knives to the finished article. He knows about verbs, pronouns, and the stuff that makes words become articles. Many times, his thoughts raise my blood pressure for a time, but as our thoughts blend to the finished product, the destination becomes more comprehensive and understandable. Thanks, Steve!

Thanks to Joe Kertzman, assistant editor of *Blade*, who smoothes my feathers, comes up with ideas about what needs to be discussed, and always has time to talk knife stuff.

Dave Kowalski was the best boss I ever had. He also made the first edition of *Knife Talk* possible and continues his friendship and support. Steve McCowen is a dreamer who shares special thoughts about knives; continues to encourage the exploration of many frontiers in the world of knives. Both are men I call friend.

Credit must go to the countless fellow knife makers from both the past and present who share thoughts through their knives that may be thousands of years old, being made today, or existing only in dreams of tomorrow. Artisans, in a timeless community bound by a common joy: the knife.

Foreword

A book, our book, yours and mine. I have journeyed through the world of knives and visited many fine people, places, and thoughts.

Dedication

To Angie and the many who have encouraged me to continue my quest or, more correctly, our quest. Clients, friends, those who simply wanted to share thoughts, and the environment I am privileged to share with nature and tools that must work.

Introduction

Welcome to the world of knives. Knife and man have developed together, from the Stone Age to today. As you read these words, man and knife continue as partners in the art of life. The knife has always been a part of the good times: the hunt, the harvest, the making of clothing and shelter. Lady knife has also been with man in times of hardship, where survival depended on planning, luck, skill, and competence. Sometimes man and knife survived, other times they failed; still they were together.

Man and knife is a relationship that is as simple or complex as the human side of the partnership wishes to become involved. The visitor to the world of knives enters of his own volition. He can choose to explore the most remote and complex frontiers or simply scan a map of the territory. The legends of Excalibur and The Iron Mistress are equally as boundless as the pursuit of the High Performance Knife. If you choose to enter the community of the world of knives, you are welcome; the rewards will be up to you. The world of knives has only one rule: Enjoy the voyage.

Contents

There are many facets of knives; no knife stands alone. While attempting to place articles in specific categories for the purpose of the book, many times we found them difficult to categorize. This is the way it should be, for the knife is a complete package. Nothing stands alone, a community, if you will. All go together.

Chapter 1: Function, Design, Techniques

Preface 8
What to Look for in a High Performance Blade – Part I 9
What to Look for in a High Performance Blade – Part II 14
Design: Linchpin of Knife Safety 18
Carbon Steel: A Proven Survivor Displays its Worth 21
The Natural Approach to Knife Maintenance 24
How to Sharpen a Knife Anywhere, Anytime 27
There are No X-Rings in the World of Knives 30
SNAP! 32
A Bear Blade Barely Meets the Test 35

Chapter 2: Knife Talk Philosophy

Preface 39
Spirit of the Knife 40
Why I Write About Knives 44
Sharp Dreams of a Frustrated Warrior 47
What About Bob? 49
Little Shop of Horrors 52
Sharpest Shop on the Prairie 55
In Search of the Real World of Knives 57
Last Scratch Fever 61
The Sheep-Horn Sherky Shuffle 64
How to Make Friends and Influence People Sheep-Horn Style 66
Dear New Knifemaker 68
What Exactly is "A Master?" 71
Mistakes I've Made 74
When First Place Isn't Necessarily the Best Place – Part I 77
When First Place Isn't Necessarily the Best Place – Part II 80
Beware of the Medicine Man 83
Guilty Until Proven Innocent 85
The Downside of Mint Condition Blades 88
Brothers in Steel 90

Chapter 3: Forging and Heat Treating

Preface 95
How to Grind the Blade 105
Tip-Top Blade Tips 111
Differential Hardening vs. Differential Tempering 114
Uncover the Secrets Hidden Beyond the Blade 116
Update: The Status of the High Performance Knife 120
Ball Bearing Steel – It Just Keeps on Cutting – 52100 Series Part 1 124
Ball Bearing Steel – It Just Keeps on Cutting – 52100 Series Part 2 127
High Performance – 52100: How to Forge It and Why It Works 130

Chapter 4: Legends

Preface 136
Paul Burke – Good Samaritan of Steel 137
Our Fair Lady of Steel 140
Paul Basch Tips on Classic Handmades 144
Myths – Ben Lilly (Break Rather than Bend?) 147
Long May She Cut! (Bowie) 151
Found: The Lewis and Clark Knife? 155
The Price Was Right – Part I 158
The Price Was Right – Part II 161
Iron Mermaid of the Great Lakes 164
Whatever Happened to Bernard Sparks 167
Last Father's Day of the 20th Century 170

Chapter 5: Outside World

Preface 173
My First BLADE Show 174
On the Oregon Cutlery Trail 177
Where Are We in the World of Knives? 181
Media Madness Is Killing Our Country 185
Dealing with Droughts in the World of Knives 188
Code of the Cutler 191
The Searchers 194
The Coming Crisis in Knife Leadership 197
Friends of the Knife 200

Chapter 1

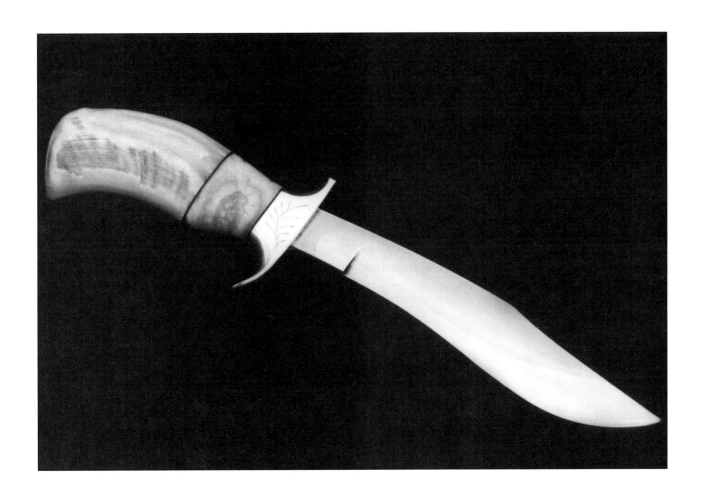

Function, Design, and Techniques

These are some of the serious and practical thoughts about using a knife, including knife function, safety, and maintenance. The high performance knife is only as good as the man who uses it. Still, the man who needs a knife can make the knife sing in his hands when they join in a true relationship dedicated to function. Some fight their partners and waste a lot of energy while others are able to work with their partner in harmony. It takes some learning, but it is all common sense once you get the hang of it. All partnerships start out through the selection process, and then grow when the right decisions, followed by practice, come together to form a team.

What To Look For In A High-Performance Knife

The rigors to which some blades were subjected convey the true meaning of high performance

Author's note: I am not trying to promote my knives by relating the events in the following, but rather to provide a complete understanding of my conception of the high-performance blade.

It seems as if a continuous battle rages between the stainless and carbon steel brothers in the world of knives. Most of the contention between the two factions arises from the expectations each has of its knives. Both sides have trouble keeping their expectations in perspective.

To many of the stainless steel clan, performance is what a knife does *not* do—that is, it does not rust too easily. They are willing to sacrifice qualities such as cut or tough for the benefits of a maintenance-free knife. There is nothing wrong with their selection of steel or expectations of performance; it is a simple matter of choice or individual preference. However, when they claim that the performance qualities that the carbon steel culture worships come under the realm of abuse, hackles rise and communication ceases. A second issue is the fre-

Taxidermist Eric Anderson with the moose rack and Fowler knife he used to cape the animal's head. He also used the knife to field dress four antelope, including splitting the pelvis of each animal and cutting the chests open by severing the costal cartilages between the ribs and sternum. He also skinned, headed, and legged the antelope.

Function, Design, and Techniques

quent and often exaggerated advertising claims concerning the "heartbreak of rust." This issue influences many in the world of knives, instilling unwarranted fear and apprehension on the part of the knife consumer. Even knifemakers yield to the fear. (There was a time when even I was ashamed to admit that a carbon steel blade could rust.)

I consider maintenance of any tool a responsibility, a privilege, an art form, a serious aspect of pride of ownership, and an opportunity for sincere participation in the nurturing of a good friend—the knife. As time and task pass, man leaves his personal signature on his knife, visible for any with vision to see long after man has passed from the partnership.

The virtues claimed for the stainless-steel knife remind me of a book written some years back. Some readers may remember the movie based on the novel, *The Stepford Wives*, in which the wives were transformed from those with free wills to absolutely obedient robots. While, on the surface, the world portrayed in *The Stepford Wives* might seem like utopia to some, life could get a little dull if that were the way it always was with knives and women.

> "When many evaluate knife performance, they are shooting a little low."

Do not get me wrong. I have nothing against stainless steel. I do not feel that it has the beauty of the carbon steel in my grandmother's silver fruit knife that it replaced, though I do feel that it has a place in civilization. Mankind has entered a new era of the disposable knife, where many who were never taught good cutlery etiquette let their blades soak in soapy water for hours rather than take three seconds to wipe them off, dry them and put them in the knife rack. Today, most kitchen knives spend their time bouncing against other blades and culinary hardware in the drawer. My grandmother was proud of her knives because they were sharp and cut well. She also took pride in her care of them. Today, many knives are judged primarily by appearance and glitter, with performance only an afterthought or theoretical assumption.

"Civilized man" is beginning to lose perspective when it comes to evaluating a knife. A lot of blades, both factory and custom, have great reputations that I feel are not warranted. Part of the problem comes from the fact that when many evaluate knife performance, they are shooting a little low. Many park bench conversations evaluate knives solely on their

These two knives were forged to be the same before being overheated. The top blade was cold forged to shape. Note the raised tip—only the cutting edge was cold forged. The blade at bottom was ground to shape. At one time, the blades were the identical shape.

10   Function, Design, and Techniques

 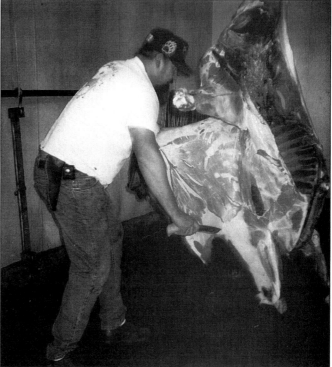

Troy Rogers begins to skin a deer. Note his grip on the handle, with his index finger in front of the guard. "He won't get 'bit' using a knife this way," Fowler observed. "Troy is a professional. He won't take a chance on using a knife without a guard." In the other photo, Troy is well into the skinning operation.

performance at the simple task of dressing out a single animal without needing sharpening. Anyone who has dressed out several hundred animals knows it does not take much of a knife to do the job. Stone-age man butchered game for thousands of years with his teeth and a sharpened rock. This does not mean that the knife that only dressed one deer without the need to be sharpened is not good enough for an individual. If this level of performance is enough for him, fine, but to call the knife a high-performance blade is an overstatement of fact. I feel that *high performance* needs to be carefully evaluated based on empirical testing, both at the time the knife is made and then in the field, in order for a rating of potential performance to be valid and reliable. I intend to discuss some aspects that I feel are required of a high-performance knife.

Functional Balance

The most important attribute of the high-performance blade is what I call *functional balance*. In the "Willow Bow Dictionary" of knife talk, this simply means the blade is designed and constructed with the greatest potential task as the guiding light.

> "I requested that he keep a record of the knife's performance."

As there are many kinds of knives, I will limit my discussion to those with which I am most familiar, namely, high-performance field knives. My conception of a field knife is one that would make man a dependable and absolutely faithful companion at work, hunting, fishing, camping, or simply seeking to enjoy being a part of his natural heritage.

It would be a monumental task for me to describe—and you to read—all the variables that need to be considered and the choices the knifemaker must make in the development of a high-performance blade. For example, years ago I decided to relate, in story form, the many issues to be considered in the design and construction of only the tip of a blade. Six thousand words later, I still had not considered all the alternatives, and the end was not in sight. I deleted the story from my computer.

There are many facets of the high-performance knife. One essential attribute is its ability to cut. Recently, I gave two blades to professionals, asking them to keep a record of how the knives performed. One of the blades went to a taxidermist, Eric Anderson, a neighbor of mine and owner of Timberline Wildlife Taxidermy. The blade I gave him was forged from a 3-inch ball bearing of 52100 steel, using my construction methods that have been covered in BLADE® many times.

Due to an error on my part, the blade was defective. After forging and normalizing my blades, I soak them at 1330°F for two hours and allow them to cool slowly to room temperature in my Paragon oven—a process I call a *bladesmith anneal*. I inadvertently soaked several blades for two hours at 1880°F—650°F hotter than planned. I polished and etched one of the blades. Etching revealed that the grain structure had grown very large, and, after this kind of "abuse," I did not know of any way to save the remaining blades. I decided to experiment with them and learn what I could.

One of my fellow ABS master smiths, Joe Szilaski, had been doing some experiments with a process called *cold forging*. Joe and I had discussed the many variables that could be involved in the process. I decided to cold forge one of my blades, then harden and temper it in my usual manner, and see how that would influence its nature. The resulting blade cut rope very well, about four times better than a second "burned" blade that was heat treated the same as the first, but without being cold forged. I was going to break it to test its strength and toughness, as well as to examine the crystalline structure of the blade, when Eric stopped by the shop for a visit. He looked at the blade and said that it was exactly what he wanted for a caping knife. He requested most emphatically that I should not break the blade. As he is a good friend, I put a guard and handle on it and gave it to him. The blade was marked "experimental, cold forged." I requested that he keep a record of the knife's performance. Recently, he gave me the first results of his testing.

Obviously, any knife that lays claim to the title of "high performance" must be able to cut, and cut well. Here Fowler uses a vintage Marble's blade of high carbon steel to make short work of a length of rope.

Test Results

Without sharpening the blade, Eric field dressed four antelope, including splitting the pelvis of each animal and cutting the chests open by severing the costal cartilages between the ribs and sternum. He also skinned, headed, and legged the antelope.

> "I ground the blade several times developing the various convex profiles I wished to test."

The next task was to cape a Wyoming bull moose. This is one job that is a supreme challenge to the cutting edge, as it requires the constant cutting between bone and moose hide that is very tenacious and hard on knives. Usually the operation requires three sharpenings on an average knife. Eric was able to cape the moose and, when finished, the edge would still slice paper efficiently. All the tasks were completed without touching the blade to a steel or stone.

As the knife is now Eric's, I cannot test the blade's toughness by flexing it to 90 degrees or more. All I can claim for the knife's performance ability is that it cuts very well. Eric asked me to sharpen it for him. In three minutes, his Norton Fine India stone and I had the knife back to the original shaving edge—which also qualifies the blade as easy to sharpen.

The second knife also was forged from a 3-inch-diameter ball bearing. A 7-inch blade, it originally was intended to become a model I call my Camp Knife. I had used it for some extremely severe cut-

"The objective of a high-performance knife is to exceed the owner's expectations."

ting tests, exploring various functional aspects of the convex grind by changing the blade's geometry. I ground the blade several times, developing the various convex profiles I wished to test. By the time the tests were completed, the blade was too thin for my version of a camp knife.

Troy Rogers owns and operates Riverton Ice and Cold Storage. During the local big-game season, he processes moose, elk, bear, deer, and antelope, and throughout the year butchers domestic American bison, cattle, sheep, and hogs. Troy is a regular visitor to the ranch. Many times I have reground the factory-made knives he uses at his plant. Troy had witnessed some of the testing I had done on the blade and literally had fallen in love with her. When I told him that the knife was destined for some destructive tests and invited him to watch, he requested that I give it to him and let him do some tests with it, using it for what it had become—a nice butcher knife. He agreed to keep records of the blade's performance in his packing plant.

After the local big-game season started, Troy worked 18 hours a day processing game. He brought his first report on the blade's performance. Without touching the cutting edge to a stone, the knife skinned, legged and headed at least 80 antelope, 27 elk, 23 deer, and eight moose for a total of 138 animals. These figures are a minimum estimate based on a carcass count. It is obvious that things were a little hectic during the 14-day test. Some other carcasses came in partially skinned, while others required some cleaning up.

Troy trimmed meat around bullet holes and used the knife for many associated tasks as he prepared the carcasses for cooling and cutting. He is a professional. He knows well the art of using and maintaining a knife. He steeled the edge using an F. Dick steel that he had customized by smoothing it out to a very fine surface with 180-grit sandpaper.

Joe Szilaski: A true mastersmith who brings a beautiful blend of traditional European craftsmanship to the world of knives is shown here testing one of his blades for strength and toughness. Needless to say, the blade passed—and more.

The knife previously had passed some extremely demanding tests requiring high levels of toughness and strength. It then proved itself beyond my expectations on wild game. In my opinion, the knife definitely qualifies as a high-performance blade in this application. While most of you who read this would never have the opportunity to work this many animals, the objective of a high-performance knife is to exceed the owner's expectations. By proving her performance qualities in marathons such as this, you can be confident she is able to serve man well.

Cut is only one aspect of the high-performance knife. There are many other issues to consider. First, the knife must be designed in such a manner that is consistent with the blade's intended use, or possess functional balance. Troy and Eric, both skilled professionals, selected knives they felt possessed the functional balance for the work they needed them to do.

What to Look for in a High-Performance Knife – Part II

Knife safety, knife use—and abuse—and much more are topics for makers to consider

Author's note: I am not trying to promote my knives by relating the events in the following, but rather to provide a complete understanding of my conception of the high-performance blade.

Any tool of man must be designed to be as safe to use as possible. I feel that a full guard is an absolute requirement on a high-performance knife. The knife I made for Troy Rogers—Troy is my friend and customer who tested one of my pieces in part one of this story—has a full guard and is therefore safe as possible. Troy's processing plant is only a mile from the hospital, but a cut hand would be very inconvenient and, at the height of game season, an economic disaster.

I feel that, for field use, a knife must be capable of any task that circumstances demand. It is impossible to abuse a knife when you are desperately trying to survive unexpected challenges. A knife that is made with what I term high performance in mind also can make life a little easier at times.

Last summer I was on a tractor baling hay. While watching the windrow—the row of cut hay—behind me, I drove the tractor too close to an irrigation ditch in the corner of the field and dropped the rear wheel into the ditch. The tractor was high centered with the draw bar on the ditch bank. I had two choices: Walk about a mile to get another tractor and a chain and pull the tractor out, or fix it with what I had. There was a pile of old fence posts on

According to the author, one of his knives was taken into a remote region on an elk hunt—led here by Mike Miller—and, after a game saw and hatchet designed especially for the job failed, was tough enough to split the vertebrae of an elk until the animal was cut completely in half.

14 Function, Design, and Techniques

> "For field use, the knife must be capable of any task that circumstances demand."

the other side of the fence that I could put under the tractor tire, dig the dirt from under the draw bar, let the tire gain traction on the pile of posts, and drive out. I used my knife to dig the dirt from under the draw bar and was back baling hay in about a half hour. The task wasn't real easy on the knife, but it made the day a little shorter and I was able to resharpen the blade in about five minutes when the day was done. I would have used a shovel if I had one but I didn't. The knife was good enough.

Fowler said that in order to be classified as "high performance," knives should be able to handle a variety of chores besides cutting. Here, in temperatures of -10°F, he uses one of his knives to chip a sheep head out of the ice in Wyoming.

Quality Control

Another critical aspect of the high-performance field knife is an absolute dedication to quality control. The knifemaker must be able to test each and every blade he makes for cut and edge flex at the time he makes it, preferably before he dedicates the time and materials into finishing the knife. It is too tempting for a maker to sell a faulty blade when he has put the time required into finishing the knife. It is best for the maker to test his blades as soon as he can, and either correct the problem or destroy the ones that do not pass muster.

I feel that there should be no secrets in the high-performance blade. It should read like a book. Etching the completed blade reveals the true nature of the steel to both the knifemaker and the client. There is no other absolutely reliable method of providing such information on each blade that is so readily available to the majority of knifemakers in their shops. By etching all his blades, the maker also will become more proficient in his heat-treating methods.

In addition to testing every blade for the basic requirements of cut and edge flex, the maker should test a representative sample of his blades to the point of destruction. I usually test one in 30 blades to the limit. While this may seem like a terrible waste, when a maker is selling a high-performance knife, he must do all in his power to assure top quality. I have not had a blade fail for some time, but the testing still goes on. This does not absolutely negate the possibility of a faulty blade leaving the maker's shop, but it does guarantee that the maker is doing his part to ensure the client's confidence in his knives.

By developing quality-control testing procedures that can be accomplished in his shop when the blade is made, the maker receives immediate feedback concerning his methods. Immediate feedback is an essential element contributing positively to the learning process and to the development of new skills.

Field-Testing For Longevity

Years ago I felt I had achieved the ultimate blade by brazing an oddball alloy onto the surface of some damascus steel I had made. Tests conducted immediately after making the blade indicated that it cut better than any damascus blade I had tested. I could not wait to show it off.

Several months later, Wayne Goddard came to my shop and I handed him the blade to test on some hemp rope. Wayne tried to cut with the blade while I anxiously awaited his accolades. After a few attempts at cutting the rope, he said, "This knife won't cut!"

"You are kidding," I shot back in disbelief.

"No, I am not kidding," he repeated.

I handed him a stone and suggested he sharpen the blade and try again. He and I repeatedly sharpened the blade and tried to cut with it. The blade

"The maker should test a representative sample of his blades to the point of destruction."

would not make one slice on the rope. After a few more attempts to make it work, I acknowledged the obvious: The knife would not cut. After much consideration, I came to the conclusion that the blade, when it was new, was truly a high-performance one. After a few months in my shop—which at the time wasn't heated—the nature of the steel had changed due to time or temperature fluctuation, or both. It was a lesson well learned.

Whenever any potentially significant changes are made in the knifemaking process, the resultant blade should be subjected to thorough testing over a period of time, use, and environmental conditions before it can be sold to a customer. As a result of this experience, I carried and used hard my first completed multiple-quench 52100 blade for some time before I offered one for sale.

Strength & Toughness

I feel that the all-around high-performance field knife must be designed and constructed in such a manner as to enable man to effectively deal with most any situation he may face in the future, whether presented by design or accident. The high-performance field blade must be tough enough that it will bend rather than break. A broken blade is of no use whatsoever. A knife that breaks has failed. Blades intended for hard use also should be strong enough that they cannot be flexed beyond a minor deviation without the added leverage of an extension to the handle. This degree of strength provides the man who may have to rely on the knife an added margin of dependability.

Last year, my attorney, who had accepted one of my knives as payment for legal fees, was hunting elk with three other men. They had packed into a remote region on horseback and set up camp. The next day they shot and killed two bull elk, but the place where the animals fell was virtually inaccessible. Fallen timber and a rock slide prevented the hunters from getting horses within a mile of their harvest. They had to hike in and dress the elk out, then quarter the bulls and pack them to the horses. They had a game saw and a hatchet with them. Unfortunately, both of these "trinkets of the big-game hunting market" failed when the hunters tried to split the first elk. My attorney—Tom is his name—decided to put my knife to the test.

The author says the knifemaker must be able to test each and every blade he makes for cutting and flexing ability—the author does the latter here—at the time he makes it. It is best for the maker to test his blades as soon as he can and either correct the problem or destroy the ones that do not pass muster.

They tied the hind feet of each elk to two trees, putting some tension on the spine. Tom held the knife inside the body cavity of one of the animals with the edge centered on the first vertebra, while his brother used a tree limb to beat on the spine of the blade that protruded outside the elk's back. One-by-one the vertebrae separated and the hunters were able to split the animal in half.

While this may sound like a Herculean feat, when you understand how the spine of an elk grows, they were cutting with the grain of the bone in the center of each vertebra. Due to the geometry of the blade, the sides of the convex blade were doing the work, splitting the bone ahead of the cutting edge. A high-performance blade is not simply ground to shape but carefully sculpted by visions of performance realities. The blade was tough enough and designed for the task.

An interesting sidelight occurred when Tom brought the knife to my shop for cleaning up after the hunt. He said that the edge had chipped! I examined

A true high-performance knife is one that performs both a variety of cuts and offers an array of hand holds.

the blade and could feel two small dents in the edge. Tom stated that he had inadvertently bounced the edge against some rocks while skinning the elk. Wanting to show off a little, I told Tom that I would fix the dents by working them out on my anvil with a small ball-peen hammer. Tom wouldn't let me do it, stating that he didn't want me to "abuse" the knife, requesting that I clean the edge up with my sharpening stone instead. I tried to convince him that the knife was born on the anvil, but he would not listen. Tom timed me while I resharpened the blade. It took eight minutes to hone out the dents and sharpen the edge. The high-performance field knife must be easy to sharpen.

Other High-Performers

Other makers seek and achieve high-performance blades. While experimenting with the multiple-quench process, Bill Burke of Salmon, Idaho, recently achieved a blade that flexed 90 degrees over 50 times, each time returning to straight. This knife is now in the hands of a butcher who is thoroughly testing its qualities cutting meat.

I have related these events only to bring understanding to what I feel are some essential attributes of a high-performance knife. Any maker can achieve these qualities in his blades, though they aren't qualities that are easily accomplished. It takes time and dedication and involves a lot of learning but, if high performance is the goal a maker wishes to achieve with his knives, it can be done.

> "Any maker can achieve these qualities in his blades, though they aren't qualities that are easily accomplished."

A friend of the late Dick Iiams hard-surfaced oil-field drill bits, bulldozer tracks, and other heavy-duty steel tools for a living. Dick's friend made a knife that could literally chop a burned-out oxygen cylinder in half. The blade was bull strong, crude as they come and would have challenged a hydraulic press in a flex contest. I judiciously used my earplugs and safety glasses while witnessing an extremely dramatic cutting demonstration that proved the man's knife was up to the task. This was his dream of a high-performance blade and he succeeded.

Any maker who wishes to make high-performance knives can achieve his goals by simply using his knives to do what he intends them to do. The high-performance knife will continue its evolution to perfection as long as the maker seeks to make the best knife for the job. Simple testing can be very revealing—just use the knives to cut what they are intended to cut. Experiment and test relentlessly. There is a lesson to be learned by every failure. Each maker must take his time and keep his mind open in order to recognize success. He should keep accurate notes and take nothing for granted. The essence of high-performance knives is not a thing to be captured, but forever sought, nurtured, and carefully groomed. Each maker should explore the frontiers of his steel thoroughly and, most important of all, enjoy the voyage.

Design: Linchpin of Knife Safety

During hunting season, the emphasis should be on knife safety—and a good place to start is with knife design

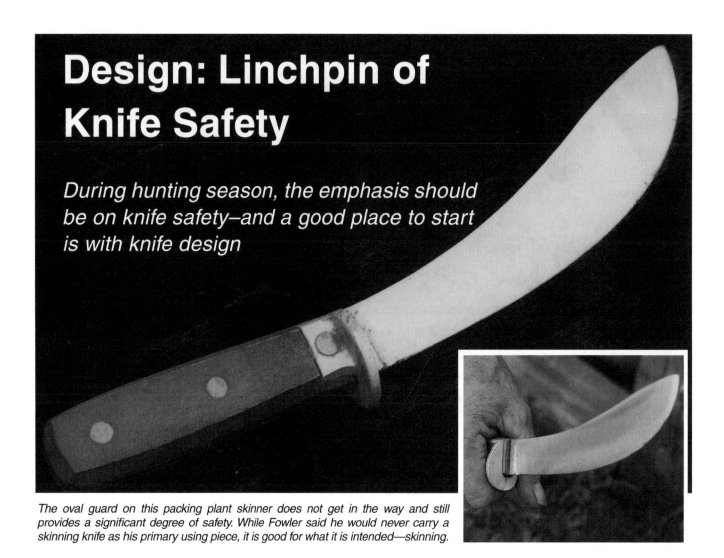

The oval guard on this packing plant skinner does not get in the way and still provides a significant degree of safety. While Fowler said he would never carry a skinning knife as his primary using piece, it is good for what it is intended—skinning.

If you are like me, you like to read and talk of heroes, those magnificent human beings who run into burning buildings, dive into freezing waters, or leap into the path of a stampede to rescue folks in danger. Heroes bring out the best in man.

But heroes risk their own lives and sometimes lose them, too. Where possible, the need for heroes must be prevented, because the events before and after the fact can be too costly. Safety through environmental design is the responsibility of all blade enthusiasts. In the world of knives, everyone involved can contribute.

While going to college, I worked many jobs, most of which contribute to the thoughts I share with you through the pages of BLADE®. During the late 1950s and early '60s, one of my part-time jobs included working for an ambulance service.

One fall day a call came in for an ambulance to meet a four-wheel-drive pickup racing to town with an injured hunter in the bed of the truck. The caller reported that the hunter had suffered a severed artery and was in bad shape.

I headed for the mountain road and had no trouble spotting the pickup in question—the dust cloud it was raising was visible long before the truck came into view. There were three men in the bed of the pickup: the injured party, another with his bloody arm extending from the pelvis of the injured man, the third working at keeping everybody calm in the middle of the hurricane of events.

I was told that the injured man had been trying to split the pelvis of a bull elk and had pulled his knife through the elk's pelvis and into the vicinity of his own pelvis, severing his femoral artery. His friends had been with him and, when the profuse hemorrhage became evident, one of them had inserted his hand into the wound and, by virtue of a lot of luck and providence, was able to stop the bleeding.

Stopping a femoral bleed of this nature is next to impossible without the aid of a skilled surgeon and a lot of equipment. A quick discussion among us reinforced our belief that we should keep with what was working, and the injured man's friend should keep hold of the artery. Since it was cold in the bed of the pickup, transfer to the ambulance was absolutely necessary for the 50-mile trip to the hospital.

With a lot of planning and coordination, we transferred the injured man into the ambulance without his friend losing control of the severed artery. The hunters were college football players, linemen—big, strong, and used to teamwork. As we headed for the hospital, above the howl of the siren our conversation was not much more strained than that of a friendly poker game, mostly aimed at maintaining the morale of the injured man and the friend who was keeping his lifeblood off the ambulance floor.

Arriving at the emergency entrance to the big city hospital, we were greeted by an obviously important and angry intern who immediately ordered the man with his hand in the wound to "get his filthy hand out of that wound!" The third hunter, I believe he must have played guard, standing about 6-foot-6 and weighing more than 300 pounds, advised the intern that "no snot-nosed twerp was going to cause his friend's death, and that nothing was going to happen until his friend was on an operating table and a competent surgeon was ready to go to work." The intern wisely left. Soon a surgeon appeared and they proceeded to the operating room, where all went well. The injured man's friend was a hero; he had not only found but held onto that femoral artery for over three hours! The muscle cramps in his forearm were considerable.

This fine old Case kitchen knife does not need a point to do the work for which it is intended.

> "He held onto the femoral artery for over three hours!"

The remedy for this kind of accident is education. Splitting the pelvis on an animal you are field dressing is not necessary. It is just as simple to core the rectum, tie it off and pull it through as you remove the intestines from the body cavity. I personally always split the pelvis when field dressing simply because an old rancher taught me how to do it the easy way. Place the knife tip on the top of the joint on the ventral surface of the pelvis, cutting edge up, with the animal lying on its back. Start about 1/2-inch back from the aspect of the pelvis farthest from you, using the knife tip as a wedge and driving it down through the joint by hitting the butt of the handle with the palm of your right hand. Repeat this action, moving the tip successively back into the unsplit portion of the pelvis until the task is complete. The spine of the blade does the work, saving the cutting edge for the rest of the job.

It is not necessary to apply any excessive force to the knife. In fact, applying excessive force to a knife is another source of accidents. Had the hunter been aware of either of these methods, the accident would not have happened.

I detest knife accidents. They can cripple and, in an instant, change the lifestyle of the injured. Whenever I become aware of a knife accident, I ask all the questions I can. How did it happen, why, and how could it have been prevented? Then, as in this case, I ponder over the events for years.

Safe By Design

While knife accidents will continue to occur, I strongly believe that makers can have a significant influence in reducing the severity and frequency of the accidents by providing knives that are designed to be as safe as possible. In addition, makers should take the time to educate those who want or need to know about the proper use of man's favorite tool, the knife. That the design of knives can and does influence the degree of risk to those who will use them has long been understood by those who carefully study issues of safety in design.

Some time ago I bought an aged bread knife made by Case from Bernard Levine. I was impressed with its design but wondered why it was made with a blunt tip. After using it for the past five years, I have realized that a sharp point is not needed on this type of knife. I use the blade in the morning slicing bread for my toast. The knife is usually stored in a hanging wall mount, about head high, next to the kitchen sink.

One morning I reached up to remove the knife from its perch. However, the blade hung up ever so slightly as I pulled it from the rack and my hand slipped from the handle, the knife falling to the floor. Not wanting the precious, antique blade to bounce on the floor, I tried to grab it in mid air, an obvious error on my part. I caught the knife but the momentum of my hand drove the blunt tip into my belly. Had Case included a sharp point on the blade, it could have cut me seriously. The blade without a point is now more greatly appreciated. Thanks, Case, for offering a bread knife, a time-honored design that does not have a point on the blade and does not need one. Sharp points are necessary for many knife tasks but not all, especially around the kitchen.

As for trying to save the knife by catching it in mid-air: It was foolish on my part, yes, but a reflex action is hard to control. That is why safety by design works. It kicks in when the mind shorts out. By using knives that are designed with safety in mind, the odds of getting through such an event are more in your favor and it just might make a difference.

Many issues of knife safety and design are not clear cut. Wrist thongs on knife handles are one such issue. A wrist thong can reduce the probability of loss of the knife when using it over water, or prevent the knife from flying across a room full of people in rope-cutting demonstrations.

A wrist thong can also pull a knife from its sheath when its user is walking through brush. Wrist thongs do not always provide a measure of safety. Wayne Goddard related one such instance.

A gentleman was walking into his kitchen carrying some pheasants in his hands, along with his knife. The wrist thong caught on the door knob and pulled the knife through his hand, cutting his hand. Personally, I do not like wrist thongs and would rather be able to get away cleanly from a knife.

Guarded Moments

I would not willingly choose to carry a knife in the field as my primary tool unless it had a full and comfortable guard to protect my hand from accidents. One afternoon I was talking knives with a man who previously had worked for many years in a packing plant skinning beeves. He showed me his hand and there was a nasty scar on the palm. His crippled fingers did not fully extend and the severity of his accident was obvious.

He told me that he had skinned cattle eight hours a day, five days a week, for over 15 years, and one day his hand slipped from the handle and he "could not get away from the knife fast enough." He was cut to the bone. Several surgeries failed to restore full function to his hand and he had to change occupations. Needless to say, he felt guards were absolutely necessary on his working knives. Still, guards can be inconvenient in many functions. Every kind of knife has its place.

> "The challenge is to keep your brain working whenever you are around any tools"

All issues of safety are matters of give and take. Anything you add to a knife takes away from another function. Therefore, there are few absolute rules concerning safety and design. All are a personal choice. The challenge is to keep your brain working whenever you are around any tools, and make decisions about design based on knowledge rather than by accident.

I fully realize that not all clients of knifemakers will want a knife that is designed as safe as possible. I do feel that knifemakers can do their part to provide as many safety features as they can in their knives, and, when talking with clients, to do their part to educate all who will listen. Also, whenever makers hear about a knife accident, they should ask as many questions as they can, determine the nature of the accident and consider any elements of design that could have prevented or lessened the accident, and they will be doing their clients a favor.

Carbon Steel: A Proven Survivor Displays its Worth

Fowler's favorite blade material may stain but that's just one of its badges of honor to him

When I read articles about how carbon steel rusts and that proclaim the virtues of stainless steel, I feel like an old coyote I once watched crossing a pasture filled with mares, young colts at their sides. Normally, a coyote would avoid such a situation at all costs but it was opening day of duck season, and the peaceful countryside was filled with hunters. Someone had shot at the coyote and he had more serious problems where he'd been than where he was going.

The first mare to spot him gave a squeal and took off after him, her warning summoning every other mare in the pasture to help defend the young colts. The pasture contained about 100 acres and at least 75 adult horses, all of the latter after the coyote. His was a spectacular display of broken-field running. They cornered him several times and he took some pretty healthy kicks from front and hind hooves, even getting tossed into the air once!

Benny Carroll, a veteran Wyoming guide and hunter, taught Fowler to smear fat from game on a carbon blade to protect the steel, here on an old Frank Richtig kitchen knife.

With his tail between his legs and all the dog speed he could muster, the coyote ducked under a fence and tried to jump a partially frozen slough. He took a long leap, not quite making it to the other side, broke through the thin ice and swam the rest of the way, parting the frozen slush as he went. He climbed up on the bank and shook most of the water off. One ear laid flat on the side of his head and he looked pretty pitiful. He glanced back at the horses, who were standing at the other side of the fence, continuing to display their outrage.

I was standing in some brush less than 20 paces from the coyote. He started to lick his wounds when he caught my scent. He stared at me and I commented that it looked like he was having a pretty tough day. He headed north and it wasn't five minutes before I heard some more shots from that direction.

The carbon steel blade, like the coyote, is a proven survivor. It has survived and will continue to survive in spite of its oft-proclaimed reputation—thanks to its stainless rivals—of a blade that rusts.

Stains As Flakes?

I was brought up with carbon steel knives and can never remember one of them rusting. I consider a stain on a blade of no greater significance than a suntan on my skin.

I always thought that the folks calling for stainless blades were kind of like the people who invented dandruff shampoo. At the time they started trying to sell the shampoo, most folks didn't really care if they had dandruff or not. As a result, dandruff-shampoo sales were pretty slow at first. Finally, advertisers began talking about the "terrible tragedy of dandruff," and folks started buying dandruff shampoo—but not before they'd been sold on the idea that dandruff was bad.

The stainless steel knives I tried during my youth failed to impress me and I went back to carbon steel. I've made a few experimental stainless blades and never achieved the performance levels that are easily within reach of the forged carbon blade. Still, many approach the carbon blade with fear and are convinced that carbon blades will self-destruct when used. I ask: How could the carbon blade have played such a significant role in the development of man and be so inferior?

I've cut steak with a carbon steel blade made by Frank Richtig, famous maker of thousands of kitchen knives, outstanding hunters, combat

Fowler said the remaining blade on his grandmother's carbon steel potato peeler is about 10 percent of what it was when new, but it still does the job.

blades for World War II GIs, and other knives, for more than 20 years. It was old and well used when I got it, and it will outlive me. I also have a knife my grandmother used as a girl. It was the first knife to cut me when I learned how to peel potatoes more than 50 years ago, and it could cut me just as well today as it did then. It will still peel potatoes with the best of them and it's more than 100 years old!

"Many are convinced that carbon blades will self-destruct."

Rust No More

Many of my customer's worry about rust and stain on carbon steel blades. At first, I couldn't consider their concern seriously until I took the time to examine the ways I've learned to care for my knives. My methods actually are simple procedures that I thought everyone knew.

I learned how to field dress big game at the ripe old age of 10. The man who taught me was the late Benny Carroll, of Sinclair, Wyo. Benny was raised on a ranch, guided and hunted all his life, and butchered his own meat. My family hunted with him every fall. (Being too young to legally hunt big game, I was the errand boy.)

One afternoon, someone gut-shot a buck antelope and wasn't too enthusiastic about dressing it out. Benny stated that if I wanted to learn how to dress out an animal, this was as good a time as any. I had my very own hunting knife and it was sharp, thanks to my grandfather's honing instructions.

The first thing Benny taught me was to smear some fat from the game on the blade and the animal fat would protect the blade during the work it had to do, as well as during storage after use. From that day on, the only time I wash my sheath knife is after some especially dirty job, such as an autopsy on a cow that's died from unknown causes or, before surgery on an animal, I'll dip the blade in a disinfectant—if one's available.

While hunting, fishing or for day-to-day work, it's the fat from the quarry or frying pan that protects my blade. Should a detailed analysis of the contents of the sheath of my carrying knife ever be conducted, the lab would find traces of fat from many species, including fish, prairie dog, rattlesnake, rabbit, raccoon, coyote, badger, bobcat, deer, elk, moose, bear, sheep, cow, and horse.

Now, I fully realize that readers in the medical profession may cringe at the thought of a pathogenic sheath. I believe that if the animal on which you're working is healthy and you find no evidence of disease while dressing and skinning it, there's little danger from the animal fat. If the danger were as great as the gossip in the daily news would have you believe, man never would have survived as a species.

From the time steel knives became available to man, fat as it came from the animals that fed him kept the steel he carried as good as the day it was forged. No mountain man carried oil with him because it was unnecessary and inconvenient. The fancy compounds now known as rust preventives weren't available, and the mountain men used what they had.

Those same miracles of nature are still available today. Whether you're fishing, hunting, camping or simply enjoying a home-cooked meal, all traditional care for the carbon steel blade is at your command. Just don't put it away wet!

Trapper Jake

To test my beliefs, I talked to my friend, Jake Korell. Jake trapped for more than 77 years. He also has a fur business and is one of the top taxidermists of all time (at least in my book). Jake has worn out more knives than most folks will ever own. I asked him if he liked stainless blades better than those of carbon steel. He didn't hesitate in his response: "I've used carbon steel knives for 77 years, skinned thousands of animals, and never had rust on one of my knives. They may stain but that does not hurt them a bit. You can't get much work out of a stainless steel blade."

> "Anyone can take care of a carbon steel blade."

One of the most famous knifemakers of recent times and I were discussing knives and steel. He stated that he would like to make carbon steel knives and push the steel to the performance limit. "The *only* reason I use stainless," he observed, "is because my customers *don't* know how to take care of a good knife."

I believe that anyone can take care of a carbon steel blade; it's just that most people have forgotten what they knew 200 years ago. Carbon steel may stain, it may grow old while the stainless blade still looks new but, to me, the perpetual "new look" lacks personality. I still find joy in looking at what the old carbon blades have to tell me. In their age, I feel there's the story of their life: honesty, a proud heritage, and most of them still able to go some more.

Editor's note: While Fowler's methods for protecting his knives from rust have worked for him for many years, if you're considering leaving animal fat on your blades after dressing game with them, in addition to the precautions espoused in the preceding, you should also consult your physician.

The Natural Approach to Knife Maintenance

From sharpening to preservation, Fowler shows you how to do it the traditional way

If you're using the knife in the kitchen, simply wash it and, without putting it down, rinse and dry it immediately. Any natural oil, mink oil, lanolin or almost any kind of wax also works well to guard a carbon steel blade from rust.

The care of your carbon steel knife is a simple matter. Men and women have lived with and used carbon steel knives handled in natural materials for hundreds of years. Many of the old knives are still around and as serviceable as they were the day they were made. A great many more have served man until there was nothing left of them.

Sometimes the relationship between man and knife is no bed of roses. The most significant problem to come to my attention is that many folks try to make sharpening too scientific. They purchase sharpening jigs and attempt to hone the cutting edge to too precise of a degree. It doesn't have be that way. I seriously doubt if anyone using a knife to do the job for which it was intended, especially a hunting or camp knife, can tell the difference between an absolutely precise, exactly centered cutting edge and one that is a few degrees off. No test that I have conducted using knives in the kitchen, on the kill floor or for field work has reflected a significant difference between minor variations in the angle of the cutting edge. If you can brush your hair, you have the dexterity to sharpen a knife that will cut very well by simply using your hands and a good stone. Stop by your local flea market, pick out a couple of those old worn knives and practice your sharpening skills until you feel you have achieved excellence. Then use the knives to cut to test the results. When you feel you're ready, go to work and enjoy fine tuning your relationship with your favorite knife.

Too Coarse Or Too Fine

All too often neglected is the value of good steel. A smooth/fine-cut combination steel will coach most edges back to hair-shaving efficiency without removing steel from the cutting edge. A lot of knives have been unnecessarily worn out by excessive use of abrasive sharpening media.

All too many stones and devices that are sold as knife sharpeners are either too coarse or too fine. If the sharpening device or stone you use is too coarse, your blade will develop a rough edge and drag against the material you're trying to cut. At the other extreme, hone a good blade on a thousand grit surface and it won't cut anything tougher than cigarette paper.

Many makers have determined the stone that works best to fine tune the cutting edge of the knives they sell. When you buy a handmade knife, ask the maker what stone he believes works best on that particular knife and use it. For most carbon steel blades, I find that the Norton fine India stone lubricated with kerosene is as good as any and better than most. A blade forged of carbon steel that has been heat treated to superior advantage usually will respond best to a sharpening grit somewhere between 240 and 320. You can even use sandpaper to sharpen your knives, with excellent results. Simply place the sandpaper on a hard, flat surface, such as a block of wood, and hone the edge to its finest potential. Free-hand sharpening on a stone is an easy skill to master. Simply work the blade as if you are trying to shave a postage stamp off the surface of the stone. Slightly roll the spine of the blade a little lower than necessary every third stroke or so, and you will avoid developing a secondary edge that is not as efficient as an edge that flows cleanly to the spine of the blade. With practice, you will gradually develop a skill of which you can be proud. Your skills, your devotion to the knife and the visions of the maker will join, and the knife will become a symbol of that partnership.

Rust Prevention By A Nose

I strongly feel that too much commotion has been raised about the problem of rust and very few discussions have taken place concerning the ease with which plain carbon steel knives easily can outlast their owners.

Rust prevention is a simple matter. If you are using the knife in the kitchen, simply wash it and, without putting it down, rinse and dry it immediately. This is the method used by my grandmother on her knives and I am still using one of them in my kitchen. As a cutting instrument, it is just as excellent today as it was 80 years ago. Anytime, anywhere you are using your knife, one of the best

The author said no test that he has conducted using knives in the kitchen, on the kill floor or for field work has reflected a significant difference between minor variations in the angle of the cutting edge. Here, he makes a detailed incision employing a choke-down grip with one of his knives.

all-time rust preventives known to man is as close as your nose. Simply rub your finger along the side of your nose, obtaining some of the natural human oils that are present there, and wipe the oil on the blade and the blade won't rust. If it is a big blade, you probably have a little more of that natural rust preventive behind your ear. The fat from the animals you hunt is also an excellent rust preventive.

Should you wish to use a commercial rust preventive, contact Mark Mrozek at Sentry Solutions and try the company's Tuf-Cloth. Follow the instructions that come with it and you won't be disappointed.

Many handmade knives come with handles of natural materials. While highly durable, such materials are, after all, natural, and therefore probably have natural enemies. Nature in her absolute wisdom perpetuates life by returning all she makes to dust for the nourishment of her "future crops." Wood, horn, bone, and ivory all need protection from the effects of mold, moisture and dehydration. Tree Wax, Mink oil, bee's wax, Extra Virgin Olive Oil, and mineral oil are all excellent products to protect natural handle materials. As noted, the natural oil from the side of your nose will also work very well to keep your investment in top shape. Again, ask the man who made the knife what he recommends.

Some locations have beetles and other insects that attack natural materials. When storing your knife for long periods in such places as storage safes, it is good insurance to place a few mothballs in the immediate vicinity of the handle. If you intend to store the knife for several years in a place such as a safety deposit box, be sure and oil the handle thoroughly. Again, mink oil, lanolin or almost any kind of wax works well. Rabbi Martin Yurman, a devotee of the world of knives, told me about some fine butcher knives that were heavily coated with Vaseline, then wrapped in butcher paper. He said they came through many years of storage in an old trunk on the humid East Coast as pristine as the day they were made.

Care of your knives is an easily mastered and enjoyable skill. It is your opportunity to build a meaningful relationship with lady knife that will be with you a long time. Don't be bashful about asking her maker about her needs and then simply share time and task with her.

> **"When you buy a handmade knife, ask the maker what stone he believes works best on that particular knife and use it."**

26 Function, Design, and Techniques

How To Sharpen A Knife Anywhere, Anytime

Always be prepared to adapt your knives to whatever situations dictate

Survival is a matter of preplanning, and knives are man's most significant survival tool. With one knife, you can make anything you need to survive. Modern man and prehistoric man have made knives that have kept the human race thriving for many thousands of years, using nothing more than what was available no matter the time or the place.

Years ago, I was in one of my favorite places, the Wyoming prairie, easily 30 miles from people, electricity, and the phone—just my horse, my dog, and me.

I've always been interested in Stone Age man and have investigated his lifestyle from whatever evidence I'm lucky enough to find. Anyhow, while out on the prairie one day I noticed a rock, originally about the size of a soccer ball, on the ground with a lot of large flakes broken from it. As there was no way that could have happened without the hand of man being involved, I took some time to look the situation over. All the evidence was there: A stone had been used as a hammer and the flakes had been knapped from the mother rock. I hobbled my horse so he could graze, loosened the saddle, and took some time carefully trying to put the rock puzzle together again.

All the pieces were there except one. I started walking around the mother rock in increasingly larger circles and, within 50 feet, I found the missing flake. The sharp edge had been retouched by the original maker as evidenced by extra chips that had been knapped off to make the flake a better cutting tool, in this case, obviously, a knife. Evidently the knapper had either killed an animal or come across one ready to eat. He'd made his knife on the spot from the materials at hand, had done what needed to be done and, rather than carry the knife with him, discarded it right where he had used it. He had no need to carry a knife with him; he could make one whenever he needed to in less time than it took me to write this paragraph. Stone Age man: competent, expedient, and at home wherever he was. Not only could he sharpen his knife anytime he needed to, he could make one wherever he was. I marked

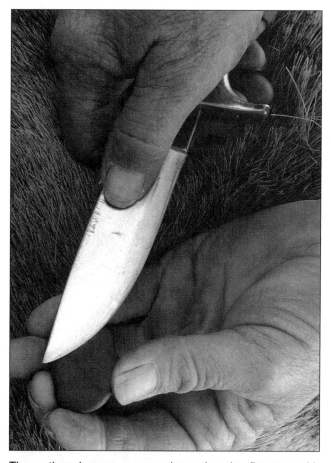

The author dresses up an edge using the flat stone his mother-in-law found for him on Cape Cod. He says he carries the stone in his pocket at all times.

Function, Design, and Techniques

One way to always have a sharpener at hand is to cut lengths from the abrasive belts used by knifemakers. Here Fowler uses a length—with the heel of his boot for a flat surface—for some edge touch-up.

the place in my mind and intended to return with my camera, but have never found the time. I apologize, as it would have made an excellent photo for this story.

This past summer a college friend of mine named Dennis Blankenbeckler and his wife, Karol, came to visit. As usual we talked hunting, fishing, dogs, and knives. He related an incident that happened while he was living in Alaska. He and a friend were hunting big game and came across an Alaskan native who had shot a black-tail buck deer. Evidently, the man had lost his knife. After some thought, he used the best tool at his disposal. He simply pounded the brass .30-30 cartridge flat that he'd used to kill the animal, then sharpened the cartridge on a stone, making a knife good enough to dress the animal out and keep from losing the meat. His comment was that it wasn't much of a knife, but it worked.

Modern man seems to have lost much of the ingenuity that his ancestors developed as a way of life. Use the tools you have, whatever they may be.

Simple Is Best

When I was asked to write an article on how to sharpen a knife anywhere, anytime, many thoughts filled my mind as to how to relate what's simple as far as sharpening is concerned. Let me elaborate.

Progress can be a trap. When man first makes a new tool, he comes to rely on it, and all the tools that took its place for many years are soon forgotten. Knife sharpeners are an example of this kind of development.

More than 45 years ago, my grandfather gave me my first fine India stone. About 15 years ago, Wayne Goddard gave me a better one. I've used the stones exclusively to sharpen all the knives that I've made or that need honing. The stones are still going strong.

> "Use the tools you have, whatever they may be."

An easy way to carry pieces of an abrasive belt for edge touch-ups is in your wallet.

I love them dearly and, when they're available, I sharpen with them exclusively. However, when they *aren't* available, it's pretty easy to improvise, providing I haven't dug myself to deep a hole.

Your ability to sharpen your knife anywhere, anytime starts with planning. Your survival system must be compatible. Should you choose to carry an exceptionally hard blade, the only sharpening tool you can use will have to be exceptionally abrasive. It's pretty tough to get an edge on some blades unless you have a diamond sharpener. I avoid such blades as if they were some kind of social disease. Why?

In the first place, exceptionally hard blades are usually too brittle to count on and, without your special sharpener available when you need it, you're out of luck should the blade get excessively dull.

One of the easiest sharpeners to use and carry is a few pieces of 220-grit sandpaper. All you need is a flat surface (I've used the heel of my boot or the buttstock of my rifle on which to lay them), and you have a pretty handy knife sharpener. The blade usually only has one or two dull spots, and it isn't necessary to wear the sandpaper out working the entire edge. Just look the edge over, find the dull spot, and touch it up. Incidentally, if you want to, carry what I think is the best sandpaper currently available. Order a Norton or 3M 2-by-72-inch, 220 grit ceramic belt from one of the knifemaker supply houses and you'll have enough field knife sharpeners to last year's. Cut the belt into 6-inch lengths and each length will fit into your billfold just like a dollar bill. Water won't hurt the belt material and it will last a long time.

When it comes to stones, Norton makes a small, fine India stone. I usually have one in the glove box of my pickup. Only trouble is, it has sharp edges and corners and is pretty tough on my pockets, so it's not likely to be with me when I'm away from my old Ford. Should the folks at Norton ever decide to make a stone about the size of a stack of three quarters with smooth, rounded edges, I'd never be without it.

Usually I can find a local stone within a few yards of my work that will touch up an edge when necessary. Just for the "long-distance company" of a man (Henry Thoreau) I greatly admire, I usually carry a small flat stone in my pocket that my mother-in-law picked up for me from a beach not too distant from the one that Thoreau talked about in his essay on Cape Cod. Many times the stone's made my cutting task a little easier. I've sharpened knife edges on cement sidewalks, rusted plow lays (the rust actually increases the abrasive quality of

The fourth hit on the mother rock with a baseball-sized stone produced a flake that would serve as a usable knife.

Necessity, the mother of invention: This .30-.30 cartridge was hammered between two rocks and sharpened on a stone to produce a serviceable knife.

the surface), the frame of a truck, and with a small, diamond-coated fingernail file (I assure you this wasn't by choice). A good knife won't self-destruct when circumstances demand using something other than the newest hi-tech sharpener.

Necessity is the mother of invention. Wherever you are, there's usually something you can use to make a knife. For instance, when it comes to field dressing animals, even a glass jar or beer bottle can be made into a knife that will get the job done. All you have to do is relax, think the situation over, and use what's available.

In conclusion, I'd strongly suggest that you practice the methods I've suggested in order to fine-tune your skills before they're necessary in the field. And remember: Watch your fingertips when you use small sharpening stones. Until you get a little practice under your belt, it's pretty easy to trim some hide.

There Are No X-Rings in the World of Knives

It's important to ask the right questions to gauge performance before you buy a blade

The X-ring, or bull's-eye, is an obvious, reliable, and valid barometer for those who want to know how any firearm or cartridge will perform. This readily available indicator of execution greatly influences quality.

On the other hand, performance is difficult to gauge when it comes to knives, especially when the consumer is looking at a variety of blades at a knife show or knife shop. Firearms, especially revolvers, have come a long way in a short time simply because they're subject to immediate and obvious performance tests. The quality cannot be disguised; guns are either reliable and accurate or they aren't. Those who demand accuracy and reliability in their firearms soon learn who makes good ones,

There are some common denominators among premium knives, too. Most significant is the attitude, dedication, and ability of the maker. Just look at all the outstanding gun makers. For example, H.M. Pope stands out as one of the all time best when it comes to accuracy. For a short time, I owned one of his homemade rifles. It was one of the sweetest shooting guns I've ever had the honor of knowing.

"What was the influence that made Pope the outstanding barrel maker of his day?" The answer is that he was a target shooter who sought to make the most accurate rifle of his time and, through his competitive interest, great intelligence, and determination, he sought and found the answers.

The same principle applies in the world of blades, though many times knife enthusiasts chase

> **"Many times knife enthusiasts chase an X-ring that's not dead center."**

an X-ring that's not dead center. For example, rope cutting is one of the best edge-holding tests that's readily available to the individual knifemaker. Many a great race has been run simply to provide the best rope-cutting edge. Still, there's a lot more needed in a high-performance hunting knife than edge holding alone. In my opinion, an edge that's made in such a manner that it will slice rope exceptionally well but won't do much else is not the edge geometry I would want on the hunter I may have to depend on if and when the chips are down.

The steel in a hunting knife needs to be able to slice rope in top fashion. The edge on each knife must have originally passed a rope-cutting test as an indication of edge holding at some point. However, the edge I carry in the field has to be able to do a whole lot more than a thin rope slicing edge can do. The edge on the hunter needs to be balanced for maximum strength, flexible to resist chipping the first time it hits bone, and tough enough that it won't break should you need a pry bar in a hurry.

Cutting or chopping wood are tasks that many knives will be asked to do, though neither is a valid indicator of edge-holding endurance. I've cut wood with a blade that wouldn't make one slice on a piece of hemp rope. Again, the best blade design for cutting wood is not necessarily the one you want to carry when going after that trophy bull of your dreams.

In other words, unlike with guns, the X-ring in knife performance is much more complex than the geographic center of a piece of paper.

Does your knife fit your hand as well as J.D. Smith's piece fits his?

Other Variables

Other significant variables in the quest for the ultimate functional knife are the individual preferences of the hand that will use it. Each man knows how he uses a blade and must seek the knife designed to complement his hand, as well as his methods.

When searching for a knife intended for use in the field, attributes such as edge geometry and overall blade and handle shape are immediately apparent. Nonetheless, the true nature of the steel is unknown until it's been put to the test. Unfortunately, the only complete test of a knife involves destroying it in a series of torturous regimens, leaving you with the knowledge that the piece would have measured up to the most demanding task but, unfortunately, with a wreck of a knife.

Technical analysis of the steel is also usually destructive and expensive. The blade made by Rick Dunkerley analyzed in "52100 From A Metallurgist's View" in the October 1997 BLADE provided what Rick and I believe to be extremely valid information as to the nature of the steel. The test cost Rick two $500 knives, one for the test and one that he traded, in typical bladesmith fashion, for the metallurgist's evaluation.

The development of this quality of knife steel took years of experimentation and elbow grease. Rick and I believed that the scientific analysis would support what we knew already due to the simple performance testing we had done. The experimenting continues and there are already indications of better blades to come.

The knowledge gained was well worth the expense, but knowledge just as valuable is available to all knifemakers by simply using the knives they make for what those blades are supposed to do.

The performance testing of knives must, by the very nature of the instrument, begin with the knifemaker. Only through his dedication can the high performance blade become available to the consumer. Makers owe it to their customers to test their blades thoroughly.

If it's a skinner, the maker should do a lot of skinning with it, not only the easy jobs like deer and elk, but some of the tough ones like beaver, moose, and buffalo. The maker must realize and inform the consumer that the knife is intended as a skinner and is not necessarily an optimum design for other tasks.

When designing a knife that's intended for "all-around use," makers must use the piece in relation to the all-around knife as they conceive it, and thus learn its strengths and weaknesses. This is the closest the maker can get to the "X-ring" of knife performance.

SNAP! That Dreaded Sound Means Disaster Has Occurred

The failure of an inferior blade is worse than having no blade at all

Sounds; they come in all sizes, from the clap of thunder to the sound of silence. Some are pleasant, while others interrupt well-laid plans. The loudest sounds of all are those paired with high expectations. The hollow click of a firing pin on a dead cartridge at the crucial moment of the hunt is a prime example.

To me, just as devastating is the snap of a knife blade in the middle of a task.

Many years ago, a fellow deputy sheriff and I were in the process of arresting one of the county's less-than-desirable felons, his hair-triggered proclivity for violence well known. My partner and I had a warrant for armed robbery from a distant

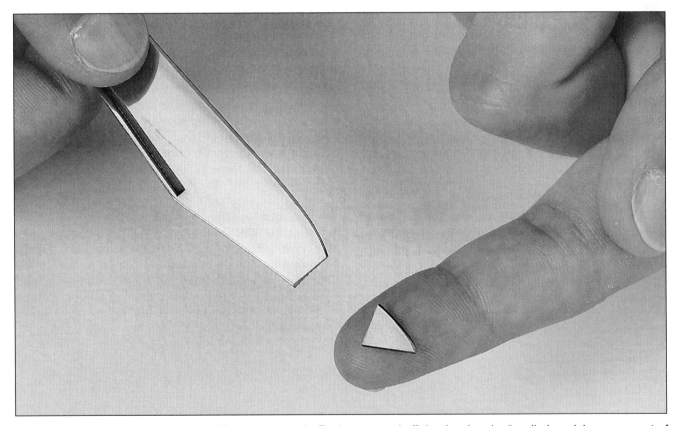

Using a pocketknife blade as a makeshift tension wrench, Fowler snapped off the tip when he "applied a minimum amount of torque" to it.

32   Function, Design, and Techniques

county on the suspect, though the felon was unaware of the warrant.

Earlier in the day I had contacted a neighbor of the suspect and requested that she call should he come home. That evening she telephoned, saying that the suspect had returned home, looked like he had been drinking and that he was alone. Wanting to take him alive and peacefully, if possible, my partner and I watched his house from a distance until he had time to settle down. We quietly approached his house on foot. Looking through the bedroom window, we could see him sleeping, a mostly empty bottle of wine at his side and a .45 ACP pistol on the table next to his bed.

The situation was perfect. If we could get to him while he was sound sleep, we very possibly would have the handcuffs on him before he could wake up. While my fellow officer watched the suspect through the bedroom window, I returned to the patrol car to get my lock picks so I could pick the front door lock.

Soon after reaching the car, I became concerned when I noticed that my tension wrench was missing from the leather pouch that held the picks. As very little torque is required to pick a lock, I decided to use my brand new, high-ticket pocketknife as a tension wrench on the keyway.

Opening the blade to 45 degrees, I inserted the tip into the keyway. I held the pocketknife between my index finger and thumb and applied a minimum amount of torque to the blade. The tip broke like a piece of glass and the pocketknife fell to the hollow, wooden porch floor with what could conservatively be described as one of the loudest sounds I have ever heard. I immediately stepped to the side, drew my revolver, and waited. After the echo subsided, I checked my fellow officer, who gave me the OK sign. I returned to the patrol car, rummaged through my briefcase, found my tension wrench, and returned to pick the lock. From that point on, all went as planned and the suspect awoke wearing my Peerless handcuffs.

The late Peter Hathaway Capstick wrote Death in the Silent Places, *in which he described the Tigero, a man who hunted big cats and the spear he used to hunt them. The spear had to be tough and could not break for the hunter to keep a big cat at bay. Here is Capstick with the piece he carried—a Swiss Army knife.*

> "I would like to see all tools of man rated for reliability at the time of purchase."

Some might claim that my use of the pocketknife as a tension wrench was an act of heresy and abuse. I strongly believe that each knife I carry should be up to any task I may need to accomplish. To me, the knife is a tool to serve man and, if it lacks the qualities expected, it is a failure. The failure of an inferior tool is worse than having no tool at all, for you depend on it and, therefore, may not have a substitute readily available. The consequences that cascade from the use of a tool that fails may be insignificant, merely inconvenient or fatal. One day I would like to see all tools of man rated for reliability at the time of purchase; this would allow the buyer to knowledgeably select a tool of the level of quality he expects.

Function, Design, and Techniques 33

Big Cat Blade

The fact that many tools, in this case knives or steel blades, have had to face extreme tests of reliability for hundreds of years is manifest in literature and gives some clues as to the challenges faced by some early blacksmiths.

I just finished reading Peter Hathaway Capstick's book, Death in the Silent Places. In one chapter he describes a Tigero (in this case, Alexander "Sasha" Siemel, 1890–1970), a man who hunts big cats—up to 400 pounds, some even larger—with a spear. The events described occurred in the late 1800s and early 1900s when big cats were considered vermin in Africa and South America. The spears used by Tigeros were designed with a cross piece forged as a part of the base of a spear blade, in this case a 14-inch one. The cross piece functioned as a stop to keep the skewered cat from running down the spear and shredding the man who held it. Once the Tigero had the cat on the end of his spear, he could keep the animal at bay and lever the double-edge blade around in the cat to do as much damage as possible. If necessary, the Tigero would withdraw the spear and lance the cat in a more lethal location.

Imagine, if you will, the strength and toughness of a blade required to withstand the leverage needed in such an event: on one end of the spear a 170-pound man, on the other a 200-pound cat, five times as strong as the man, each of the combatants fighting to the death with all the power, skill, and adrenaline at his command. The blacksmith who made the spear had to know what he was doing. Should the spear break, the cat might suffer a lingering death later, but the Tigero would have known the full fury of the cat and, in all probability, died almost immediately. As most of this kind of hunting was done in dense brush, tall grass, and afoot, the spear could not be too heavy, not only to conserve the strength of the Tigero, but it had to be light and balanced in order to be quick—in this case, quicker than a cat.

Modern man lives in an age where many weapons and tools are born and nurtured by high technology. No longer does man's national defenses depend on the skills of the 18th-century blacksmith. Plowshares will not be forged into missiles, aircraft or computers. People tend to look to technology and science for solutions to the challenges they face. Technology and science do well on the technical stuff, but when it comes down to one of man's most basic and useful tools, the knife, modern man has not surpassed what the blacksmith of yesterday in many countries achieved—the knife absolutely devoted to function.

> "Plowshares will not be forged into missiles, aircraft or computers."

The handmade knives of today know many benefits that were unknown to the knives of yesterday. There are a large variety of steels from which to choose, and they are better than ever. A vast degree of technical information is available at man's fingertips or the local library, a great advantage over the village blacksmith hundreds of years ago who had to discover most of his techniques solely on his own. Today, many willingly share information instantaneously with other knifemakers thousands of miles away, to the benefit of the knife itself, the skills of the individual craftsman, and the man who wants or needs the cream-of-the-crop of contemporary handmade knives. The individual bladesmith of today is limited only by his dedication, curiosity, imagination, and skill.

A Bear Blade Barely Meets the Test

Don't let inexperience and "a piece that's popular" get you in a bad situation

The time was the late 1950s. I was going to college, working at a dairy in exchange for room and board. One Saturday morning an elderly neighbor drove his pickup to the milk barn and asked if I could help him with a bear he had shot the day before. As he was both neighbor and friend, the owner of the dairy said it was OK. If I missed that afternoon's milking, he could handle it.

I picked up a jacket and my elderly neighbor/friend and I headed for the hills. As he drove, he explained that he'd been deer hunting along a deep canyon and had spotted a bear on the other side. He shot at the bear and was positive that he'd hit the animal hard. The bear then had scampered uphill into some scrub oak and my friend didn't see it come out. Due to the remote location and the time it would've taken him to get to the bear, he decided to wait until morning.

By the way, any hunter knows that an animal mortally wounded will go downhill rather than uphill. One of the absolute rules of hunting big game is whenever you approach a wounded animal or one that you believe is dead, ALWAYS approach it from the uphill side. NEVER approach it from below, as a wounded animal will always run downhill without regard for what or who is in the way.

Anyhow, I'd seen my friend shoot on other occasions and strongly felt that the chance of him hitting a bear at anything but close range was remote at best. Being neighborly, I kept my doubts to myself and felt that it was a good day for spending some time in the hills, as well as an excuse to get away from the dairy cows for my first day off in some time.

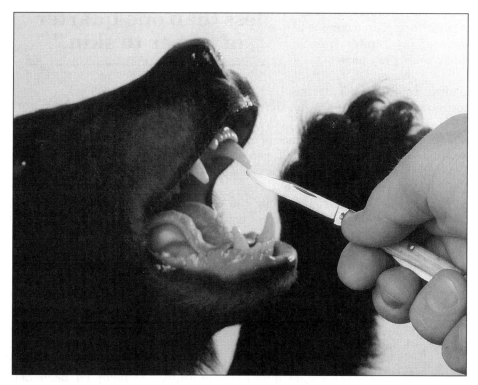
If Fowler had his druthers, he would've had more than a penknife to dress out a 250-lb. bear.

Function, Design, and Techniques

My friend drove to the place where he'd shot at the bear. The canyon wall on our side was nearly straight up and down, a long way down! He pointed out the brush where the bear had disappeared, a patch of scrub oak about 50 acres in size. The only way into the canyon was about three miles downstream. The plan was that he would drive me to the trail leading into the canyon, then drive back to the place from where he had shot. I would walk upstream under his direction from the canyon rim, locate the bear and skin it out, preserving the claws and skull intact. I would then pack the trophy back to the trail and my friend would buy me dinner.

It was about that time that one of the hunting magazines had published an article claiming that you could always tell the tenderfoot hunters by the size of knife they carried. The larger the knife, the less experienced the hunter, or so the author claimed. I knew better but, for a lack of confidence in my beliefs, had been carrying only a small penknife for some time. There was a kind of competition among many hunters, the man with the smallest knife being the winner. I was real proud of the fact that I could dress out a deer with a small pocketknife to the amazement of some of my hunter friends, thus proving my prowess.

As I descended the trail leading to the creek at the bottom of the canyon, I began to feel a little inadequate with the insignificant pocketknife I carried. But what the heck, the distance of my friend's shot would have been more than 300 yards. He couldn't shoot that well, therefore, there probably was no bear to skin anyway. I knew that!

Going was a little tough along the creek that had made the canyon, but in several hours I was at the right place. My friend directed me to the spot where he'd last seen the bear. I began to get a little concerned when I saw bear tracks on the ground, along with blood, hair, and stomach contents, indicating a gut-shot bear. Contrary to usual wounded animal behavior, the bear seemed to have headed uphill into a tangle of scrub oak so thick that the only way I could get around in it was on my hands and knees.

I knew there was a wounded bear in that jungle of brush, and the neck muscles that make the hair on my neck stand up were beginning to ache. The more time passed, the more I missed having my custom Smith & Wesson .45 revolver with me. I tried carrying a club but the density of the brush rendered that idea impractical.

Leaves were falling from the oak and tracking was next to impossible. After much searching I found the bear. Fortunately, he'd died the night before. He'd picked a good place to die, tangled up in scrub oak, making the next few hours as difficult as possible.

Skinnin' Time

I started skinning the bear. Again, all the blade I had with me was the dinky little penknife.

The bear wasn't large. It weighed about 250 pounds and was fat. Most folks don't like to eat bear meat. I've eaten worse but the bear's being dead as long as it had been made the decision to leave the meat an easy one. The best cookies I'd ever tasted were made with bear fat and I would've liked to have taken some of the fat with me. However, the distance and terrain dictated tough going. And as I didn't want to carry any more than was absolutely necessary, I skinned the animal as tight as possible, leaving most of the fat on the carcass.

> "This left me with one tiny, itty-bitty penblade about 1/2 inch long, and a little less than one-quarter of a bear to skin."

Tile little knife rapidly began to leave its impression on my hand. The spine of the blade cut into my finger while the heraldic worked on my palm. I cut a strip from the tail of my work shirt and wrapped as much of the knife as possible to provide some relief for my hand. As I started to cut off one bear paw, I applied too much pressure and the clip blade of the penknife broke cleanly at the ricasso.

This left me with one tiny, itty-bitty pen blade about 1/2 inch long, and a little less than one-quarter of a bear to skin. I carefully continued work on the animal, taking time to sharpen the tiny blade with a river rock, hoping against hope that the blade would outlast the bear. A job that should have taken an hour kept me busy most of the afternoon.

Moonlight guided me back up the trail to the canyon rim. I'd worn out my 5 a.m. breakfast about 16 hours before the bearskin was in the pickup and my friend and I were headed for dinner.

From the time I started down the trail to the bear, I'd been debating the wisdom of itty-bitty, tiny knives as an indicator of prowess in the woods. Each man must make his own decision as to the size of knife he carries. I've irrevocably, totally, and absolutely concluded that no matter what, I'll never consider myself properly dressed for any occasion, except those that dictate a suit and tie (I don't even own one), unless I'm carrying the largest knife I can.

I showed my friend what was left of the knife that had skinned the bear. He said, "Why didn't you say something?" I was rather reluctant to admit that I didn't figure there was a bear to skin, so I simply said that my pocketknife had been "good enough," considering my skill with it. He proceeded to show me several fixed blades he had in the pickup and one on his side. "These knives aren't much but they are a darn sight better than the one you used," he said. "Only a man bent on self abuse would try to skin a bear with something like that!"

Conclusion

As we headed home, I looked at what was left of my itty-bitty pocketknife, holding her in my worn-out hand. I spent a little time cherishing the memories of the events we'd shared and then tossed the knife out the window. While I wish I had it to photograph for this story; one just like it will have to do. All you have to imagine is the clip blade gone and the pen much shorter and thinner to know what she looked like. I don't miss her very much. I think I learned my lesson.

At the time I was young, not too smart, and believed bad information when I should've known better. I took too much for granted, put myself in a situation that could very well have been deadly and wasn't prepared for what could have been. A lot of "ifs" didn't happen—some did. The only harm was some blisters from that itty-bitty, tiny knife, a little hunger, and a lot of fatigue. Looking back on it all, I was lucky that I had the opportunity to learn and, hopefully, not let "a knife that's popular" get me in a bad situation next time.

Chapter 2

Knife Talk Philosophy

This section, for want of a better title, addresses the environment where knife, man, woman, and community meet. An attitude or thought where there may be no absolute right or wrong. Tradition lives here, as do dreams of beauty, honor, and individual creative expression. Here are thoughts that many have known, but fail to talk about. Here lives the background in the landscape of the world of knives. It is a place of fun and sometimes a little aggravation, but all the same an essential part of the world of knives.

Spirit of the Knife

Determining exactly what that spirit is lies in the heart of the maker

Some feel that abilities such as singing, acting or, in this case, making knives, come to the individual as a gift. Some talents may be gifts but they require a lot of support to develop into more than "what could have been." When motivation for a certain endeavor is high, individuals who initially seem to possess very little talent can overcome the odds and contribute greatly to their fields of choice.

Many have heard a child's first tune on a piano, and I've seen the first knives from the hands of knifemakers. Some reveal the seed of greatness, and subtle hints of originality speak to the individual's potential contribution. If there is a gift, surely it exists blended in the nature of the individual as a talent, ability or desire to take it further. It's a talent awaiting nurture both in the form of environmental influence—family, peers, lifestyle—and opportunity, as well as from within in the form of curiosity, desire and, to some extent, intelligence.

I've seen the first knives of makers that manifest true creativity. Then, in the pieces that follow, despite that first resourceful yet crude knife, the maker's inspiration is lost to the standards of tradition or shortcuts that lead to mediocrity. The maker at one time had a gift but lost it, his talent or genius wasted in his choosing to follow the ruts of tradition. The potential imaginative contributions of the individual remain forever dormant. He who is simply given certain talents without personal conviction and stimulation to fully develop his dream most likely will take them for granted and fail to develop his gift to full potential.

I've received several great gifts. One is good health. Another is an insatiable curiosity in some domains that interest me. The third was a very special grandmother. Her gifts were numerous. She instilled in me the gift of absolute faith and trust in Our Creator, including a love of all the experiences He provides, with a special place for the exceptionally honest beauty of nature. Grandmother had no enemies, forgave every transgression, and thus was free to enjoy a very special and complete love for all God's creations. She was extremely talented musically. She sang, played the piano and organ, and willingly taught all who wanted to learn. Her talent may have started as a gift but her ability grew throughout her life. She was a lady who believed in absolute, unshakable commitment and taught all of us who were able to listen that should we desire anything special, we had to work at it for it to be worthwhile, and that any task could and should be enjoyed. Such commitment is the essential element that makes progress and sanity possible. Setbacks or failures were actually opportunities most beneficial when we chose to consider them as a time for learning.

So it should be with everyone. Nothing worthwhile in life is free, but the ability to enjoy the quest and to explore

Some makers will select sophisticated tools to aid them in their quest of the knife with spirit, while others will develop their skills with hand tools and knowledge. Errett Callahan used stones to knap this dagger from one solid piece of emerald green glass. Overall length: 11-/12 inches. (Weyer photo)

The essential element of the knife with spirit is the degree of involvement and emotion the maker devotes to its construction while always involving that special part of himself which will identify the knife as coming from his hand. This mother-of-pearl folder comes from the hand of Barry Gallagher and includes the maker's "Bug Splats" damascus blade steel. Closed length: 4 3/4 inches. (PointSeven photo)

with enthusiasm the frontiers of one's love provides an opportunity for good times.

Nickel Cigar Knife

Grandmother knew my interests were different from hers and she never tried to push me into a career in music. She was there to teach but I wasn't there to learn—at least not music.

As my interests began to grow, she encouraged me without hesitation. I gave her one of my first knives. It possessed all the grace and dignity of a nickel cigar. She praised it highly and years later requested one with a slightly thinner blade for kitchen work. Never did she have anything bad to say about that first knife. She was always using it when she knew I was about to visit. She used the knife only to encourage me.

If any gift awaits the artist, it's the reward he receives through self-satisfaction as he develops his talent, enhanced by the pleasure of sharing the emotions that surround the knife with others. When the knifemaker lives with his craft and is truly involved, his spirit lives with the knife. The object of his labor receives his full awareness. For a time, the knife and maker are one. However, when the knife is finished the maker parts with it, usually without regret.

This facet of knifemaking for some time eluded me. I felt that somehow I should hold on to the individual knives after they were finished or feel regret when they went to new homes. In some way, I felt I had to be remiss in my duty to not feel loss when my knives "left the nest." While the creations of the knifemaking craft may remain in the maker's heart for a time, for most when the knives find a new home, it's a simple fond farewell. As with my grandmother's songs, once sung they were gone, but the melody continues for those who truly listened.

Seeking to explore the emotions and thoughts surrounding my relationship with the knives I make, several thoughts come to mind. It seems that I make each knife in the faith that it will be a final resting place for my devotion. As a result, I seek to instill my best ability in accord with the nature of the individual piece. When the knife is finished, my emotion moves directly to the next object of my attention. As I make knives, I seek to impart the special qualities of all previous pieces I've known into the next. Therefore, no knife I make really leaves my shop.

It's the same with each maker. As his talent grows, he'll find a time that the many subtle events he shares with the knife are of a nature that even those who are very close to him won't be able to see or hear. While he doesn't wish to keep them private, the sharing of these events known only to the maker and the knife becomes unlikely because his communication with others may be lacking. The maker can smile as his enthusiasm resides in the knife, like a time capsule awaiting the day another craftsman will hear the message he and the knife have to share. These gifts and more unite to become the true piece made by the knifemaker.

A Special Quality

For some time there has been discussion concerning what exactly is a knife made by hand. I don't wish to enter into this debate but simply want to discuss a special quality, a quality that I call spirit. The knife that manifests the spirit that somehow transcends other knives is the result of a true relationship of devotion, commitment, and respect among knifemaker, knife, and client. I feel there can be no rules to define it, no rules to delineate a specific level of quality. The answer lies in the heart of the maker.

The youth who makes that first piece reaches a few simple decisions as he starts his journey into knifemaking. At first the decisions are simple; use the metal that's available, shape it into something that cuts, put a handle on it, "and smile." It's the degree of his involvement in the knife that loudly proclaims the inclusion of the maker's spirit in the object of his attention. The knife with spirit—or spirit of the knife—is as important to him as any Loveless, Moran or Scagel. At this point in the maker's development, support is critical. His fam-

It's Fowler's opinion that the most significant attribute of the knife with spirit is the number of decisions the maker reaches as he nurtures the piece to completion. Jack Levin's sidelock folder features frames carved from a solid piece of damascus in a baroque style. The fluted damascus blade was inspired by those in a 19th-century collection of fine European knives. Closed length: 4 3/4 inches. (PointSeven photo)

ily or friends may like the knife and, if all goes well, praise his accomplishment. He has now tasted both the joy of success and has completed his first knife with spirit.

The essential element of the true knife with spirit is that it's based on a firsthand emotion received not from tradition but from the life of the maker augmented by his freedom to involve his true self in the knife. The quality of the knife depends on the degree of freedom, involvement, and emotion the maker devotes to its construction, while always involving that special part of himself which will identify the piece as coming from his hand.

Quality comes with time. For the maker who continues his journey into knifemaking, he will further develop not only his talents but will seek more knowledge as long as opportunity and curiosity encourage him to search, question, and experiment. The more he learns, the more decisions he makes while developing the knife. His decisions affect both the nature of his knives as well as their quality.

While others will develop their skills with hand tools and knowledge, the most significant attribute of the spirit of the knife is the number of decisions the maker personally reaches as he nurtures the knife to completion. The nature of the involvement is multidimensional, a blend of his total life experience combined with the opportunity to develop pieces that can be recognized as being more than just knives. Each piece is different but recognizable as coming from his hand. As with the animal species, knives made by hand and with spirit develop as if guided by some Darwinian process.

Some make knives that have both beauty and function by simply setting up a machine and, for one time only, being intimately involved with the piece. The first knife made in such a process is in

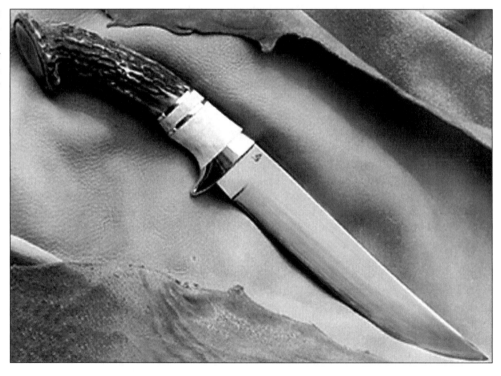

In Fowler's opinion Joe Szilaski is a true master, his dreams in steel bring top European craftsmanship to the world of knives. He is a true friend who shares thoughts on a regular basis.

When the knifemaker lives with his craft and is truly involved, Fowler writes, his spirit lives with the knife. Lloyd Pendleton's spirit has been coming through in his blades—such as his stag hunter—for almost 30 years now. (PointSeven photo)

According to Fowler, there can be no rules to define a specific level of quality for a knife with spirit. The answer lies with the maker. Matt Diskin's answer lies in a mosaic damascus "sommelier's" folder with corkscrew. The handle inlay is elephant bark ivory. Closed length: 4.5 inches. (PointSeven photo)

the spirit of being made by hand but those that follow, at least in my opinion, come into the realm of production. If the knives the maker and machine provide create emotion in those who see or use them, they qualify as art but aren't quite as special to me as the one-of-a-kind knife. Thus, in my opinion, such pieces leave the realm of being made by hand and with spirit.

Some makers may push the machines they know well to the absolute limit of their potential. There are also artists of a special nature who dance with advanced technology. They love to create a very special level of the knife with spirit. Ray and Ron Appleton are the best examples. The maker who works with his hands, knowledge, and state-of-the-art technology also reaches many individual decisions concerning the construction of the knife, most of which will be known only to him. Such artisans impart knowledge, emotion, and true love into each one-of-a-kind knife that comes from their shops. They are deeply involved with the product of their attention throughout every moment of its development. They incorporate their own essence, blending self with material and machine to develop that gem of man called art—the unique, hi-tech knife made by hand and with spirit.

The atmosphere—which includes the total environment, steel, maker, design, and more—that surrounds the spirit of the knife made by hand reflects the degree of involvement that I believe is another and most essential aspect of such a piece. Next comes the knowledge developed in that pursuit. Finally, the harvest of all that came before is invested in one single knife. There's a great difference between the knifemakers who fashion pieces for a living as opposed to those who make knives and through them live a life many would envy. The spirit of the knife made by hand from those who choose to be a part of their product transcends the common piece made by those who somehow remain outside.

Knife Talk Philosophy

Why I Write About Knives

I'm a knifemaker and also a knife writer. While the two go together pretty well, they're not without conflict. In fact, I'm often asked, "Should a knifemaker write about knives or concentrate on knifemaking, leaving the writing to the knife writers?"

I write about blades for many reasons. Writing about them actually started as a result of making them. Potential clients wrote and asked me about knives. I had to answer their questions, and it soon became obvious that the best way to answer most of the questions was through writing a booklet. I also wanted to answer questions that don't get asked but should.

As I wrote the booklet, I became more and more involved not only in writing about knives, but also in making them. In my case, the two complemented each other to such an extent that they became one. If the old question, "Which came first, the chicken or the egg?" applies, in this instance I think it was a case of two being conceived as one.

My motives for writing about knives are many and varied. One is the obvious fact that many knife writers, as well as highly competent writers from other fields visiting the world of knives, while not lacking in interest, intelligence or good intentions, are first and foremost writers, and actually know very little about knives. They write interesting stories and take excellent photos. However, they're easily misled by the information they receive. Many misconceptions are printed, as happens in all fields, and are repeated by other writers and slowly become accepted as fact. As a knifemaker, I found myself wondering where reality was. Sometimes, the information made sense to the point that I actually thought that the recommendations the writers made were valid. However, when I tried to duplicate the recommendations in my shop or in the field, their folly became apparent.

If I, as a knifemaker who should know better, had to work the ideas over for a time before I could

One reason the author writes about knives is to discuss how well they perform. Here, Jouni Kellokoski demonstrates the performance of one of his Kellam Tommi knives. Kellam's address: 1770 Motor Parkway 2A, Dept. BL, Hauppauge, NY 11788, phone: (302) 996-3386.

Bill Burke demonstrates some of the heavy lifting that blends with creativity in the making of knives.

come to a conclusion as to their validity, where did that leave the average knife enthusiast? As Wayne Goddard so aptly stated, "There are very few islands of reality in a sea of fantasy and misconceptions."

I don't mean to condemn all knife writers. There are some writers who have been writing for some time whom I admire. They've written on a wide range of subjects, providing food for thought for many of us. Their written word has introduced many individuals to the cutlery community, and the world of knives has grown and continues to welcome many enthusiasts. I, for one, probably wouldn't be writing these words if it weren't for their contributions.

The simple fact is that I write about knives because I believe in them, the stories they tell and the part they've played in the survival and quality of man's existence. My greatest curiosity is in the performance aspects of knives. Only dependable knives that cut well hold my attention for long. Though a particular aspect of workmanship or design may capture my fancy for a time, the knife as a whole may lack any redeeming qualities as far as use.

Often I wonder why designs and materials that totally lack knife function continue to thrive. It isn't that I can't hear another artist's message. I'm many times touched when knives from times past come to my attention when specific aspects of design speak of a special dedication from their maker. I do, however, believe that if it's a knife or pretends to be a one, it should either cut well or be labeled, "For Ornamental Purposes Only."

The Tradition Trap

I don't claim and never hope to know all there is to know about knives, function, and steel. Those who do outlive all hope for the new frontiers. Tradition is a trap; it may demand craftsmanship but only follows the feint scent of creativity. Tradition buries the soul and is the mire from which the only escape is curiosity and enthusiasm. Though tradition is at times both profitable and safe, those who live within its ruts will progress little toward new horizons. The more knowledge gained, the greater the foundations for the future.

Many times my thoughts lead me one way and tradition pulls the other. Truth is often without the safety net of tradition and leaves one alone. Still, trying to defend tradition over what I believe to be the truth is much more costly, as the cost is my soul.

As a writer, I write about knives and dreams of knives as they come to me. With a little effort, I hope that I may be able to help bring an understanding of what knives are all about to those who read. The expectations of knives to be are much greater than the achievements that have been. Those who thirst for the knives of the future, while appreciating the dreams of the past, know the true joys the world of knives has to offer. While some will hold to the traditions of the past without understanding their nature, they cling to an empty shell and lay in the dust behind the bladesmiths of vision. The inquiring minds of the world of knives stand today upon the frontier, as I hope will all future knifemakers, providing nourishment for the dreamers and explorers of tomorrow.

"Why do I write about knives?" Because I can't help it!

Sharp Dreams of a Frustrated Warrior

The mere thought of knives offers the author relief from today's "civilized" battles

In the past, the author has criticized the numerous factory bowies of the 1800s—this one by J. Lingard of Sheffield, England—feeling they served no purpose other than the dreams of man hopelessly trapped in civilization. "Now I can fathom why the unsung heroes of yesterday loved those knives," he writes.
(The Antique Bowie Knife Book photo)

Man fights many "civilized" battles, from those of his youth, to the streets of the city, to the riots involving thousands of people, including the personal tragedies of individuals ravaged by alcohol and drugs. When occurring on a large scale, such conflicts may be observed in the daily papers, though most are noted by only the few involved. Some of the battles challenge certain individuals, and heroes are born.

Events that afford man the opportunity to place himself in situations where his well being is at stake—and quite possibly his life or that of another—cause the adrenaline to flow. There have been many attracted to these events simply for the sake of the challenge and the experience of the dynamics themselves. Others meet the challenge simply because they must. The parallel drawn between the lives of Crazy Horse and Gen. George Armstrong Custer by author Steven Ambrose is but one example.

For the serious battles, the high-performance tools of man totally dedicated to function know a home and may spell the difference between an event that makes headlines or simply results in a moment's excitement and failure.

There are those who live with nature close up and personal who

Knife Talk Philosophy 47

witness flood, drought, angry, sick or hurt animals, mud, snow, freezing temperatures, and desert heat. Much of the time they're alone and must depend on their equipment to get them through the task at hand. Again, the call is issued for superior tools of man that must meet the challenge of the moment, for quality may make a difference.

There's another much greater battlefield, dignified by Henry David Thoreau as that of men who "live lives of quiet desperation," which affects many more people. On it occur the everyday events that require extreme bravery and perseverance that are rarely dignified by glory or require any technological support. These are the routine, civilized battles where an incompetent or unscrupulous lawyer, boss, co-worker or civil servant, with the stroke of a pen or unjust shake of his head, can start a cascade of events that may well jeopardize all a man or a family has spent a lifetime building. These are the daily battles of civilization where the heroes are unsung. They won't make the papers or the daily news but the participants are heroes all the same. They fight the creditors when they are swindled and cannot pay their debts. They fight the insurance companies, the traffic jams, the crowded supermarkets, the battle of the bulging waistline, and the all-night barking of the neighbor's dog.

Conflict of Little Consequence

Today, I fight one of those obscure battles of little consequence and seek comfort. The new-age computer that I'm told is logical and used with great proficiency by my neighbor's daughter who, with pacifier in mouth, sits before it and negotiates through complex programs, is a mystery to me. At least my computer doesn't make sense to me and I don't know enough about what I don't know to be able to ask meaningful questions. Where there once were words in my DOS program, now there are pictures of something much too small to be consequential or recognizable that purport to do something. When I move the mouse to one of the pictures and click the button, the computer may simply do nothing or send what I was working on somewhere else.

I would rather fight nose-to-nose to the death with an enraged grizzly than know the absolute frustration and failure I face when trying to negotiate through Windows 98. My old DOS word-processing program worked just fine. My 386 computer was my friend. I liked it. But soon, I'm told, even my latest computer won't be able to communicate with "the new stuff." I can't receive pictures on it and have to rely on others to keep me informed of what sometimes may be matters of true consequence.

Knives To The Rescue!

Many times I've criticized the numerous factory bowies of the 1800s and the Rambo stuff of today as trinkets, feeling they served no purpose other than the dreams of man hopelessly trapped in civilization. Today I must change my opinion, for simply looking at photos of these historical trinkets allows me to experience a few moments of tranquility as I seek to escape from the frustration of the new computer. Now I can fathom why the unsung heroes of yesterday, trapped in civilization, loved those knives. The battles they had to fight didn't require knives that cut, flexed, pried or even felt good in the hand. All those knives had to do was provide moral support and enhance the dreams of the frustrated warrior on the battlefields of legend where heroes reigned.

Civilized Battle of Steel

One of the "civilized" battles that affects knifemakers is the scale that forms on blades when it's necessary to heat them above scale temperature. I've experienced a few personal battles wrapping blades in foil to prevent carbon loss at high temperatures. Paragon Industries has introduced an argon induction unit for its ovens. It's an affordable installation that hooks up to an existing unit or can be included when ordering a new oven. It's simple to use, easy to hook up, and inexpensive to operate. The presence of an argon atmosphere assures a friendly atmosphere for blade steels at high temperatures, which sometimes is necessary when heat treating or just plain experimenting. Read the instructions and use the induction unit in a well-ventilated area. Thanks, Paragon!

Paragon's address: 2011 South Town East, Dept. BL3, Mesquite, TX 75149-1122, phone: (800) 876-4328.

What About Bob?

When one partner does not support the other's enthusiasm for a hobby or interest, something special is lost

As upstanding citizens of the world of knives, many of us have special interests and, of course, many of us love knives. If you are reading this and do not like knives, it may be because your spouse, son or daughter who does like knives read this essay and respectfully suggested that you read it as well.

Do not blame me; it is not my fault. I am only the messenger, relating incidents as they happened. Only the names have been changed in the following to protect those who need protection.

As knife lovers, we are especially fortunate when our spouses also love knives or support our enthusiasm for blades. I have been married several times and have been privileged to view the subject from both sides of the fence, as well as from the standpoint of straddling the fence. Presently, I am very fortunate that my bride shares my enthusiasm for knives and other interests. She does not get angry when a new book, knife or firearm shows up, and even occasionally helps me reload. I have experienced both men and ladies who ordered knives for their spouses and significant others, working an extra job to come up with the money. I feel they were the lucky ones as they unselfishly shared each other's joy.

As knife enthusiasts, sometimes it happens in our lives that our partners do not share our enthusiasm for blades or our other hobbies. I have known men who faced what mildly could be referred to as absolute and total adversity when they fell in love with a knife. At times, some of us, just by way of fate, end up with wives or mothers who have no appreciation, understanding or sympathy for the man who wants a knife—or another knife, as the case may be.

I believe that most cutlers and many citizens in the world of knives know well what it means when a man who wants to purchase a blade requests that we send no mail to his home, or leave any messages on his answering machine, concerning the knife he has ordered. Once I was even asked to simply hang up if a female answered the phone. In my files, the cards bearing the names of such men are red tagged and bear the inscription, "wife hostile." Incidentally, and fortunately, these cards are vastly outnumbered by "wife friendly" or no notation at all.

Not long after Angela and I were married, she mailed a knife to a client who anxiously had been awaiting it. Several days later, Angela received a phone call from an irate

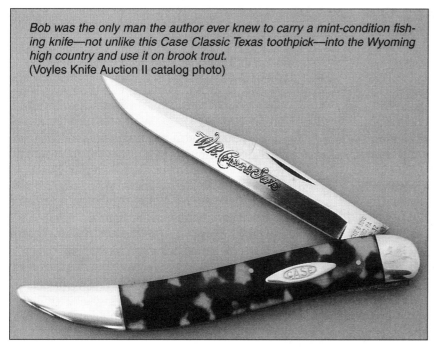

Bob was the only man the author ever knew to carry a mint-condition fishing knife—not unlike this Case Classic Texas toothpick—into the Wyoming high country and use it on brook trout.
(Voyles Knife Auction II catalog photo)

An example of a partner's shared enthusiasm for a special hobby or interest is Renza Sewell (left), the No. 1 backer of her husband Logan (right) and his love of antique bowies. At center is ABS master smith Harvey Dean admiring Logan's Searles/Fitzpatrick classic at the 1999 BLADE Show.

lady wanting to know why we sent her husband a knife. The lady claimed, with emphasis, that her husband did not order the knife and knew nothing about why it came to their post office box. Fortunately, Angela caught on and stated that maybe we had made a mistake and the woman should simply return the knife. Angela thanked her for being honest and informing us of our mistake. The knife returned in two days, the package showing a little "smoke damage," though otherwise in good shape.

Good Ol' Bob

The ultimate event of a similar nature occurred recently to my friend, Bob. He was one of the finest men I ever have been privileged to know. He was also the only man I ever knew to carry a mint-condition, mother-of-pearl-handle fishing knife into the Wyoming high country and use it on brook trout.

Bob both loved and provided for his wife in style, was always good natured, and was the kind of guy you could count on to keep his word. He was absolutely dedicated to the study of knives. He knew factory and custom blades thoroughly. Bob could pick investment knives and make an honest dollar thanks to his extensive knowledge.

On the other hand, Bob's wife did not share in his love of knives. Anytime Bob showed an interest in a knife, his wife was well known for her immediate, vocal, and forceful attitude. In deference to her wishes, Bob kept his knife hobby remote from her scrutiny.

Bob had another love. His most prized single possession was his first car. He had bought it while in high school and kept it in mint running condition for over 40 years. Since his wife did not share in his love of the car, it was never allowed in the garage and he was unable to keep the paint as pristine as the other features of the car that were protected from the weather. Apparently, his wife was afraid her Mercedes would catch sticky valves or contract "paint fade" from Bob's treasure.

While an engine overhaul went unnoticed, a new paint job would have had Bob living in the doghouse for months. His car was not anything special to anybody but him and those who love old dependable vehicles that can go for untold miles without the plastic falling off. His beloved car did not have air conditioning or any of that fancy stuff; it was just something to get a guy around. His wife did not ride in his vehicle—ever.

Through the years Bob kept up with the knife market, trading blades on a regular basis and investing his earnings in a collection of some of the finest and rarest pocketknives known to the world of cutlery. To keep from upsetting his beloved bride, he kept the knives in the trunk of his automobile in several old suitcases he had customized to afford the pieces an environment where they could be well preserved.

One day fate pointed her finger and Bob died before his time. Several days after the funeral, Bob's wife called Bob's best friend, Al, and told him to come and get Bob's "piece of junk" out of their backyard. She gave Al the title and told him that Bob wanted him to have the car. She refused any payment and requested that she never see the car again.

Al drove Bob's car home and started to clean it up, planning on giving it the paint job that Bob had wanted and bringing the fine old vehicle back to the beauty she once enjoyed. As he cleaned the car, Al noted three large suitcases in the trunk. Up to that point, no one but Bob knew about the treasure chests. Friends had seen the knives but only one or two at a time. Al opened the suitcases, discovered their treasure and immediately called another friend of Bob's who was also a knife collector. As they went through the collection, they realized that Bob had done very well trading in knives.

After much consideration and discussion, they went to visit Bob's widow intending to talk knives. They were met at the door by a lady who very clearly proclaimed that she just wanted to remember Bob and not be bothered with "the rubbish" he called friends. They left.

Being honest men and true friends of Bob, they were faced with a real dilemma.

Bob was their friend.

Bob loved his wife.

Bob had never wanted her to know about his knives.

She believed that Bob had honored her request to refrain from trading in knives.

She had, what some might call, fond memories of Bob. If they told her about the knives, her memory of Bob would forever be that of a devious spouse. Thanks to Bob's other investments and estate planning, she was very well fixed and did not need the money. Bob had no children or family other than his wife. Bob had requested that the beloved car be given to his friend, Al. Obviously, Bob did not want his wife to know about the knives but had left the solution to the situation up to Al and friends.

> "He was the only man I knew to carry a mint-condition, mother-of-pearl fishing knife into the Wyoming high country."

After much thought and consideration, the decision was made. Al and his friend sold the knives, keeping only a few special ones for themselves and to subsidize a paint job for Bob's old car. The rest of the money was donated anonymously to Bob's favorite charity.

When we share, we all benefit. I have found that true knife lovers can be trusted, and, as long as their wives, mothers or girlfriends are not hostile, we can share the good times openly. To deny another's dream only means you will not get to share the joy that dream could have been for two.

Little Shop of Horrors

Should a knife workshop be spic-and-span or "lived in"?

By Mrs. Angie Fowler

There are moments we all remember with "crystal clarity." I find this quite interesting, as most of these "life moments" seem to meld into a large mosaic, where all the colors blend into one another, the edges no longer defined. Some moments, however, are recalled with sharpness. They may have occurred decades ago, yet they're remembered as if they happened only seconds before.

My own moments of crystal clarity have been experiences of either great pain or great pleasure. The first time I laid eyes on Ed Fowler's knives was one of the latter. *Those knives!* Though I knew absolutely nothing about the "world of knives" at the time (my coworker, John McNeil, had just got me interested in knives, and I literally did not know there was such a thing as a knife show), my soul told me that Ed's show table was filled with the

Somehow, the vision of Ed making knives while Angie cleans up behind him just doesn't get it.

52 Knife Talk Philosophy

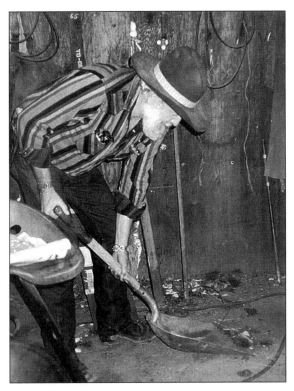

The ol' shovel is Ed's favorite cleaning tool.

Among the visitors to Ed's shop are his trusted Labs and an occasional lamb or three.

extraordinary. Those blades (and what I now know to be temper lines) screamed out to me. The beauty of the handles, the buttery colors of the sheep horn, each with its distinctive look, echoed from the rafters. There's *nothing* in this world like *those knives*, and that feeling continues to this day. I'm continually awed by the beauty that Ed's hands create. It mystifies me still and excites me to the bone. I still scream wow! every time I gaze at them—*every time!* Little did I know then that, one day, I'd get to live where all those knives are born.

By the way, I'd known Ed for over a year, actually. We'd written each other many times and had talked on the phone for more hours than I could count. He'd encouraged me to fly to Wyoming to see his ranch. I'd grown up on Long Island and had never been out west, and it sounded like a great adventure to me. After much encouragement, he persuaded me to visit.

At the end of 1995, after working up the courage—I never was much of a traveler, I finally decided to fly to Wyoming. (Believe it or not, after three days we were married, and the rest is history.) I got to see the very shop in which *those knives* are created. I remember the first time I laid eyes on that shop—actually, the entire ranch gives the phrase "the land time forgot" new meaning.

Of Mice And Man

I see the shop about 50 feet from the house. It appears much smaller than I had imagined. I walk up to the shop door, open it, and peer inside. This is a "crystal moment" as I think and say out loud, "*This* is where those magnificent knives are born? Incredible!"

> "The dust—
> the mice!"
> *Angie Fowler*

This is a nightmare! The confusion, the chaos, the sheep horn strewn about, the pieces of worn-out belts all over the workbench—at least what you can see of it. There are tools, bits of paper, magazines, old Wendy's hamburger wrappers, pencils, invoices from 1990, tubes of Super Glue and food—which Ed claims is only more insulation for his drafty little shop—*all over* that bench! The dust—the *mice!* (Ed encourages the mice because he says their nests are also a great source of insulation for the shop walls.) This cannot be real. The shop is more than disorderly. It seems like what's left over after a tornado blows through. This cannot be where those magnificent knives are born!

The impact of this "crystal moment" might not have been so startling had I not worked in New York hospital laboratories for 20 years. Precision, cleanliness, and order reign supreme there. To have it otherwise would spell disaster.

I'm also quite orderly by nature; anything else knocks me off balance. To say the least, I had no balance left after 30 seconds in Ed's shop!

I try to regain my composure, act impressed, and hide my near disgust. It isn't easy. The most outrageous part of it all is that Ed knows where everything is! He actually makes order out of the chaos. He encourages the mice to nest by tossing his crumbs in the corner. And the mice aren't the only visitors. He has the ever present Labradors: Buck, Bonnie and Red; Chamberlain, the cat; and an occasional calf or lamb. Last summer, there was even Bobby the Bat, who decided to live in the shop doorway for a week. All are welcome—*none* are turned away. Ed doesn't need to sweep; his footprints clear a path. He doesn't need to straighten up because his well-being is upset by order. In all the clutter, he finds some kind of solace—and he keeps producing *those fabulous knives.*

I made the mistake—*once*—of "straightening up" for Ed. I wanted to help out and, though he never said a word, it slowed his knife production down by one-third. I know better now and just leave him be. I realize that the shop is his world and he's agreed to leave the house to me. I'm impressed by how this quite mesmerizing man can make order from chaos and produce such consistently terrific work. I've decided not to interfere with success—but he'd better not try to set up shop in the house!

> **"Now I love the man as much as I do those knives, but I still can't believe they're born in the chaos before my eyes."**
> *Angie Fowler*

Sharpest Shop on the Prairie

Angie likes a clean shop, whereas Ed, who prefers the "lived-in" look, is allergic to "cleaning days"

By Ed Fowler

Whenever I read about another knifemaker and the story starts out about how immaculate his shop is—comments like, "you can eat off the floor" or "each machine sparkles like it's brand new"—I get to feeling kind of guilty. I never knew exactly why I felt this way until the night I started writing this story. As with all my stories, it slowly developed while I worked in my shop, making knives. I searched the deep recesses of my mind and tried to discover the source of my guilt. I think I've started to unravel the answer.

While it's impossible to recall all the events that relate to my deep-seated sources of anxiety, some come to mind.

The first

Enter your own Mr. Ed joke here.

dates back to when I was about 5 years old. The city was doing some work on the street near my house. My friends and I had escaped the confines of our yards and were watching the men and equipment at work. As they moved down the street, they left behind an absolute treasure. They'd spilled some tar!

Ed Fowler takes a dim view of his wife's efforts at cleanliness.

Knife Talk Philosophy   55

As the tar cooled, it was like clay. I decided to make a gingerbread man out of it. I always liked gingerbread men and hated to eat them because then they were gone! I'd hide them just to have company when I was confined to my room for some transgression. Whenever my mother found them, they got thrown out for being stale, attracting mice or being nothing but crumbs. I felt that if I made one from tar I could keep it forever, and so my project began.

I had a good friend in those days. His name was David. He and I worked on many great projects together. We were the kids with whom other mothers didn't want their kids playing.

David decided we should make a tar snowman. This seemed reasonable and we started putting one together. Trouble was, the snowman kept melting down as we made him. Realizing that the tar snowman project would have to wait for cooler weather, David and I made a tar gingerbread man. We soon found that if we kept dirt on our hands, the tar wouldn't stick too much, and it also helped keep the gingerbread man together. Had events been a little different, this could've been the beginning of two of the greatest sculptures the world had ever known—pure art from the unadulterated minds of creative geniuses!

We would've been OK but, before we knew it, time had slipped away and we could hear our mothers' voices calling from the real world. Anxiously we ran to their call, not as fast as we could have due to the increased size and weight of our shoes and clothing as a result of all the caked-on tar. I can still see the excitement in my mother's eyes when I showed her the tar gingerbread man that would never get stale, attract mice or break into crumbs. She looked kind of like my cat when my dog was too close to the kittens.

I tried to explain but her voice was much louder than mine. One thing good about the deal was that she didn't touch me. She just gave orders about taking off my clothes and putting them in the garbage can. I hid the tar gingerbread man on the back porch while she was looking for something to remove the tar from my skin, which was the inspiration for an event to be discussed over and over again for years to come.

Smell Of Success

On the basis of the aforementioned incident and several hundred similar ones that occurred later in my life, I've come to realize that I'm the type to become totally immersed in my work. The surroundings matter very little. The smell of a decaying buck sheep's head goes completely unnoticed when I see the beauty in the sheep horn that adorns the skull.

Knifemaker Ray Appleton came to my shop once. Upon entering, his comment still charms my ears: "Nice shop; my kind of place for making knives." We worked on several projects, then needed the contribution of a specialized machine. We visited a machine shop that sparkled. Ray's observation: "Come on. We can't work in a place like this." My sentiments exactly. There are places to work and there are places to keep clean. I have to admit there are times it gets so cluttered in my shop that there isn't room for the dogs, cat, and me, and something has to be done. However, the whole time I'm picking things up, my mind is a long way off, working on the knife that will come when the picking up is done. A vacuum cleaner always sounds angry to me, and I'm uncomfortable in shops that sparkle.

> **A vacuum cleaner always sounds angry to me, and I'm uncomfortable in shops that sparkle.**

My shop is not immaculate. While I could eat off the floor, most folks wouldn't. I have mice, dogs, and a cat for company most of the time. They don't complain about the shape of my shop and we're all more than comfortable. My power hammers are covered with 20-years accumulation of grease. The walls have never been washed or painted, as many important phone numbers and messages are written on them and are where they need to be. When the floor gets cleaned, it's with a scoop shovel. My shop broom is 20 years old and has another 20 years left in it. This isn't something of which I'm proud, nor do I apologize for its condition. My shop is for making knives, pulling porcupine quills out of dogs, warming up newborn calves that get chilled down, skinning out a deer, or whatever else we decide to do there.

The thought that a clean shop speaks to the quality of knives made therein is like judging the accuracy of a rifle by the condition of its stock. They may or may not go together.

My shop is my world. It's where I work, where I dream, where I seek the knife that will be made tomorrow, the place wherein all the sheep horn, ball bearings, tools, and machines from the past and I come together looking toward the future.

In Search of the Real World of Knives

The unsung heroes of the real world of knives are the foot soldiers who support the army of knifemakers

By Angie Fowler

This "world of knives" is filled with such interesting folks. There are genius-quality people making the best knives ever. It always awes me how there are more and more folks wanting to learn the craft—and not just men, anymore—and there is just so much to do!

I'm very lucky to be married to one of those creative geniuses—Ed Fowler—the lover of 52100 ball-bearing steel, sheep horn, and dedication to the functional knife. Ed is about as much of a genius as you can get. Sometimes he seems to be somewhere else—not even on the planet! "Somewhere out there" is where all his great articles come together, and where all those ideas about improving his knives germinate. It can almost be too spooky to be around.

I don't know if you folks can identify with what I'm trying to say here. Creativity is something very special, and artists are very special—and you know what else? The folks in the background behind these very special artists are very special, too. They are the unsung heroes of the world of knives.

I don't want to offend anyone but I feel that it's time that some light shines on all the "other heroes" of this great world of knives. I speak of the significant others, the wives, the friends of all these creative knifemakers, yes, even myself, Ed's "present wife" and oftentimes "pain in the ___"—especially when I do *not* agree with something that the highly creative knifemaker is attempting to do concerning his craft. We unsung heroes are in the background, sometimes never even seen or recognized in public. We are the ones who quietly promote the knifemaker/genius. How? By answering the mail; bringing in another income; paying medical insurance; helping with bills; keeping the books; reminding the awesome knifemaker exactly who it is that he's speaking to when his memory fails him; keeping track of all the records; making sure the money is in a safe place at the shows; trying to stabilize his environment so he can continue with all his creativity; etc., etc., etc.!

Think about it for a minute; when these knifemaking folks decide to dedicate themselves to their art, it's pretty rough going at the start. They can

Angie Fowler: a foot soldier supporting the army of knifemakers.

be like the starving artists in the movies. It takes awhile to establish themselves, especially if they don't have the ultimate benefit of having some "big name" to help them. Yes, you admire them, but you also must admire the unsung heroes behind the scenes who keep the household going in the lean times.

We unsung heroes are the ones who get the new ideas bounced off us; who have to understand when the "art" comes first; who have to appreciate the "artistic temperament"; who have to hold down the 9-to-5 jobs; who have to accept the fact that the phone will ring night and day; and that the customer *does* come first.

I once made the dumb comment to Ed that after all this time setting all his own hours I thought he would have a difficult time holding down a *real job*—hence, this article. Now, for what I *meant*: In the sometimes boring "real world," we unsung heroes have to live our lives by the clock. Just having to be *at work* at an exact time takes some discipline, after all. We unsung heroes have to answer to someone else. That's reality. Lots of folks have to do it. What many folks tend to overlook is that there's a beauty in that discipline, too! We unsung heroes are *not* necessarily the boring ones at all. We are the soldiers. May I also be so bold as to say that without the soldiers, there would be no army of knifemakers!

The payoff for the unsung heroes is that we enable those creative geniuses to blossom. We help those folks do what they love so very much—make knives. We get to see what their hands can make, and we know that we are partly responsible, though perhaps not nearly as creative.

So ... when all you great folks go to those great knife shows and handle all those great knives, *please* remember that somewhere there's someone in the background helping that creative genius you're talking to produce his art. We may not be as creative, but we are very necessary.

To all you unsung heroes—salute!

In her "real life," Angela Rapone Fowler is a medical technologist at Riverton Memorial Hospital in Riverton, Wyo.

The author's visions and the desires of his clients dictate his real world of knives

By Ed Fowler

Most of my time is spent on a private island: my place. I came to the Willow Bow Ranch outside Riverton, Wyo., nearly 30 years ago and, in some minor short-term ways, may have shaped it to some degree. The impression the Willow Bow Ranch has made on me, however, is much more significant. It is a land of knives, dogs, horses, cattle, crops, books, and whatever else nature or I feel is appropriate.

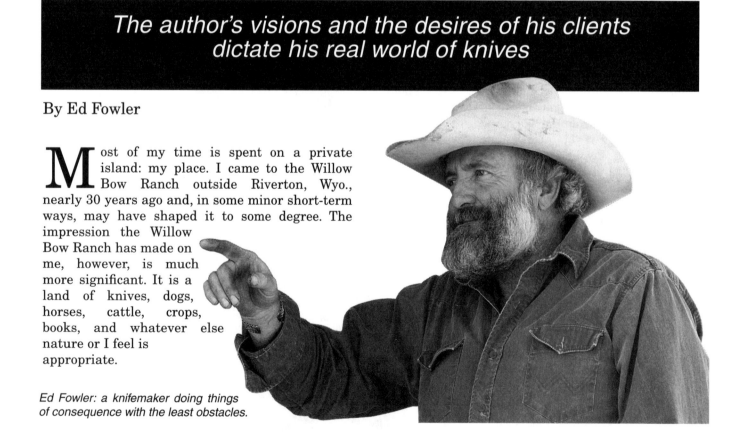

Ed Fowler: a knifemaker doing things of consequence with the least obstacles.

"Sometimes, by fate, I'm forced by circumstances beyond my control to enter into 'the real world.'"

Most of the time, I remain contentedly isolated from what my bride, Angie, refers to as "the real world." I haven't taken the time to watch, read or hear a newscast since what's-his-name made his second campaign speech for president more than eight years ago. I must concede that much of my freedom is possible through my bride's devotion to the stuff I don't do well.

Sometimes, by fate, I'm forced by circumstances beyond my control to enter into "the real world." The cognitive process that laid the foundation for this discussion started while—thanks to a kidney stone—I was filling out admission forms at a local hospital. When I came to the "name of employer," I wrote "none." The lady at the admissions desk asked me if I had a job. I said, "No, I don't have a job." Angie immediately changed my answer, stating I was "self-employed." The thoughts that follow have accompanied me while irrigating the fields, branding calves, and fooling around with knives.

I find something lacking in what comes along with a sure paycheck or what my "present wife" calls "a real job." I can only remember having one "real job" in my life. As a youth, I worked for a husband-and-wife team who constantly fought and kept me in the middle—always. One afternoon I quit without even going back for what would have been my first paycheck. It's not that I haven't been employed; I've received many paychecks since that time, but only for doing what I would have gladly done for free anyway, and felt pretty fortunate that some folks were willing to pay me for "just being me." My life habit has been that when what I'm doing becomes a job, I simply leave and find another place.

The past 20 years our crops have been planted, fertilized, irrigated, and harvested on time, and always have provided high yields. Last year we

Ed, here at work in his shop with his good friends on the Willow Bow Ranch, says "his place" in the real world of knives is a land of knives, dogs, horses, cattle, crops, books, and whatever else nature or he feels is appropriate.

Knife Talk Philosophy 59

"I find something lacking in what comes along with what my 'present wife' calls 'a real job'"

weaned a 97 percent calf crop and sold the calves to our advantage at the right time of the market.

Throughout the greater part of my life, I've devoted my leisure time to the study of the knife as a tool of man. I've done my best to investigate and promote the development of the forged blade, and hope that I've been able to influence the world of knives in such a manner that some of the blades being made today are a little better than those made during the past 300 years or so.

I feel very fortunate to have the world of knives included as a highly significant part of "my place." In the world of knives, I'm free. Time travel is available through a single knife made hundreds of years ago—or yesterday. When I share a few hours with an old knife, I'm instantly transported back in time as my thoughts mingle with the creativity of the man who made her. I don't need to travel great distances, for, in a way, my "destinations" come to me; all I have to do is see and hear the message they bear. As the message blends with my thoughts, all is complete. I find no desire to be any other place, for I can achieve a complete and honest peace of mind here, my place, be it at the forge, sharing time with a piece of sheep horn, a tree, blade of grass, or a coyote making his rounds.

Unlike those unfortunate folks who have real jobs, I never have weekends off. On the other hand, when someone tells me "have a nice weekend," it's kind of a wake-up call to be grateful that I haven't had to watch the calendar or clock.

My watch rarely, if ever, is close to the correct time. I never know the correct date and sometimes have to wonder what month it is. I don't come home after work, put my feet up, and turn on the TV. I don't get a paid vacation or have to worry about accrued sick time. If I don't work, I don't get paid, and someone else must do what needs doing in my place. I go to my tasks when they need attention, be it a cow in trouble having a calf at 3 a.m., or irrigating at first light or the last light of day. Nature dictates my workday on the ranch. I enjoy working with her for she is an honest partner. We share time and every moment is filled with the art of living. When I fail to heed her call, a calf may

As notes Angie Fowler—here at husband Ed's side at a past BLADE Show West—"when all of you great folks go to those great knife shows and handle all of those great knives, please remember that somewhere there's someone in the background helping that creative genius you're talking to produce his art."

"I don't need to travel great distances, for, in a way, my 'destinations' come to me."

die, or the crop may be short for lack of water. I'm reminded of my failure by the mournful call of a cow for her departed calf, or as I walk past plants that have suffered for lack of nourishment.

My visions and the desires of my clients dictate the nature of my knifemaking time. *BLADE®* does provide me with a deadline, and sometimes I'm late or early, but never derelict. I've never taken a vacation that didn't further my knowledge, or read a book that didn't give me some thoughts to share with *BLADE* readers. To me at least, the real world is what we awake from, when and if we seek or accept the true frontier that quietly awaits our attention.

I guess the best way to put it: Being unemployed allows me to do the things that I feel are of consequence with the least obstacles.

Last Scratch Fever

The driving force for knifemakers can be almost too small for the untrained eye to see

I've been fooling with knives for more than 50 years. Most of that time I've been vaguely aware of a single driving force common to all knifemakers, one that pushes them to the limits of their talents. I've searched for this common aspect in all cultures of the world of knives, examining each maker's product in anticipation of discovery, only to be disappointed time and time again.

Could the driving force be that the knife is man's most essential tool, providing him with food and other tools since his creation, which has in turn instilled in him the innate need to make knives? Such an explanation, while seemingly plausible, still doesn't reveal why all men aren't obsessed with knifemaking, nor explain why many makers are into knives that serve only the dreams of man.

Could the driving force be the obsession with design or materials from which some knifemakers seem to have no hope of escape? No. Again, many knifemakers only see design and materials as a means to an end, the end being a knife that they like.

Could the driving force be that makers are compelled to build knives simply to escape the boredom of television or the 40-hour work week? Again, most makers put in many more than 40 hours a week working on knives. Most of their lives (or at least mine) are spent alone with dogs or cats or mice for company, and still they seek more time in the shop.

Finally, after years of searching for an answer, I feel that I've discovered the driving

With his knife clamped in a vise, the author uses a loupe and magnifying glass in search of the last scratch.

> "In the Fowler dictionary of knife talk, this is known as 'The Syndrome of the Last Scratch.'"

The author said most of his life is spent alone in his shop with dogs, cats or mice for company, yet he continues to seek more time there.

force common to all knifemakers of all time and all cultures, as well as the many craftsmen who have graced mankind since time began. The answer was there right in front of my eyes, from the first knife I ever made to the last one I worked on today. It's obvious and well known to all. My revelation?

The Constant Challenge

The one aspect of knives and knifemakers that's known to everyone in the trade, that driving force which plagues them constantly and pushes them to the upper limits of their abilities, is a constant challenge from making the first knife to making the last.

All knifemakers know it well. When they made that first knife, it was more significant then. As they progress in ability and their eyes become more talented, they still find that, while smaller in size, it's even more challenging and, as its size diminishes, the motivation to remove it becomes even more demanding.

What is it? Simple! It's the last scratch! When each maker made that first knife, the last scratch was probably so big that you could lose a dime in it. Then, as each maker progresses, the last scratch becomes harder and harder to see and, while the last scratch gets smaller, the motivation (obsession) to remove it becomes greater.

For the bladesmith, it starts with the deepest hammer mark to be ground out of the blade. As the smith works the blade down, it's the last 36-grit scratch, then the last 60-grit scratch, and on down to the last 1,000-grit scratch that drives the smith to perfection.

When it comes to finishing the knife, that last scratch keeps the smith devoted to the knife for hours. How simple the lives of bladesmiths would be if they weren't tied to that last scratch.

The last scratch doesn't even have to be a scratch. It's the search for the best that the last scratch demands that's so important.

The "$100 Knife"

One time not long ago, I decided that I'd like to make a simple knife that I could sell for $100 or less. I started to grind a blade from flat stock, but decided that I couldn't be content with any blade that wasn't forged.

I fired up the forge and put some flat stock in it to heat up. Then I decided that if I were going to forge a blade, it had to be 52100 from a ball bearing.

I put a bearing in to heat up and, as I forged it, I tried to decide on an easy blade that wouldn't take long to forge. Well, there wasn't any sense in not making the blade as good as I could, so I put more time into its construction.

I figured that I could make up the lost time by shortening the heat-treating process. Still, I didn't feel that would be appropriate. Therefore, I kept

"It starts with the deepest hammer mark."

On his "$100 knife," Fowler figured that he could make up lost time by shortening the heat-treating process. However, he didn't feel that would be appropriate, so he kept the heat-treating time the same as for his standard knives.

the heat-treating time the same as for my standard knives. What the heck, I thought, I could make up the time and not finish the blade like I usually do.

As I tried to find a place to quit sanding the blade, again I was caught by that last scratch. The blade ended up with a mirror polish. Since it was polished, I might as well go ahead and etch it. I could make up the time lost with a simple handle.

When it came time for the simple handle, I was just going to fit up a piece of wood from an old shovel handle and epoxy it to the blade. Still, it wouldn't be as well-fitted without the benefit of a cap between the blade and the handle to keep out moisture. I started to fit a simple cap out of an old strip of copper, then decided that since I'd come this far, I might as well go ahead and fit a brass guard on the tang and merely affix it with Super Glue. Again, since the guard was fitted so well, I might as well go ahead and silver solder it to the blade. By this time, the knife was too nice for the old shovel handle, so a sheep-horn handle it had to be!

You probably have figured it out by now. The $100 knife became a $300 knife on which I probably lost money trying to make it cheaper.

In the Fowler dictionary of knife talk, this is known as "The Syndrome Of The Last Scratch." It starts ever so simple, one big scratch, followed by smaller and smaller scratches that become much more significant than that first, big, last scratch. Soon the individual is trapped and becomes a knifemaker or craftsman, not by virtue of any altruistic motivation, but simply because of that last scratch, forever caught by the last scratch syndrome. Knifemakers are driven to high levels of achievement by something of which they want to rid themselves: the last scratch!

The Sheep-Horn Sherky Shuffle

His knife handle of choice takes the author on an adventure only he could have

Making knife handles out of sheep horn has introduced me to many folks and situations that have been interesting, to say the least. Natural materials come from Mother Nature and, while she has her laws, when man gets into the equation, all kinds of things happen.

When I first started ranching, the Riverton Livestock sale barn used to do business in a lot of old horned buck sheep. Some nights hundreds of the aged bucks came through the sale ring. I always was amazed by the majesty and beauty of the horns the old bucks carried on their heads. There aren't many folks north of the border who care to eat the sheep, so most of the animals end up in Texas, then across the border into Mexico. In the early days an old buck would sell pretty cheap, especially one that was obviously on his last legs and unable to make the trip to Mexico.

I decided to try making something out of the horns and bought a few of the animals. As I learned more about them and the knives I made were getting easier to sell, I bought more sheep. The ranchers watched me buy the old bucks and, as most of the ranchers didn't know me, stories started to circulate as to my need of the critters. Some said I ate them; others just thought I was kind of eccentric.

One afternoon in the sale barn I was sitting next to an older gentleman. We were talking cows and sheep as the animals came through the ring. An aged, dying buck plodded in and I bought him for the grand sum of $1. The man asked what in the heck I was going to do with the old buck. I showed him a knife and told him that I made knives and used the horns for handles. He said, "Heck, I go to lots of sales and can pick some up for you."

I figured it wouldn't hurt to have a few extra sheep and explained that I wanted old, sick, and dying bucks with good horns for as little cash as possible, and I would chip in for gas to help him bring them to Riverton.

About a month later, I got an early call from the sale barn. They said that they were unloading some bucks for me. It was an exceptionally cold, snowy morning with the thermometer at minus-35 F. I put the stock racks on my pickup and headed for the sale barn, figuring on three or four old bucks. As I drove into the lot, all I saw was a semi backed up to the ramp. I started to get kind of suspicious. Looking close, I saw that they were unloading sheep. Looking a little closer, my fears were confirmed. The animals were old horned buck sheep. I later learned that the man I had been talking to was one of the

biggest-order buyers around when it comes to sheep. A few to him meant less than a thousand.

The driver was apologetic as many of the bucks had died on the trip, but he said his insurance would cover them. My bill still came to about $400 more than I had in the bank. This was one time in my life that I was very thankful for a blizzard!

I told him to just pile the dead ones up in the corner of the parking lot and I would haul them away for him. He was grateful and I was thankful. I thought about trading him the dead ones for the live ones, but he probably wouldn't have gone for that.

It took me most of the day hauling the live bucks home seven at a time in my pickup. I hired a kid to help me load the dead ones, and we piled them up in a far corner of the ranch for the coyotes and whatever else would eat them, figuring to pick the horns up in the spring. I had enough sheep horn to last years.

Deep In Sheep

The live bucks presented a different kind of opportunity. I knew of no way to take the horns off a live buck without fatal consequences. While I was trying to find a way to use the meat after slaughtering them for their horns, the ones awaiting a final destination continued to eat hay--lots of hay.

My part-time hired hand refused to enter the corral, a lesson he'd learned when he was a kid. His folks had raised lots of sheep and he had the pleasure of graining the bucks. Walk into a pen of old bucks with a bucket of grain and they get pretty competitive for their share, and things can get interesting. A little kid walking through a bunch of bucks with a bucket of grain learns a lot about broken-field running. I tried to find a way to eat the meat, as it was getting more expensive to feed the sheep all the time. Meanwhile, their value in the sale barn had hit rock bottom. Old buck sheep have a very distinctive flavor, kind of like liver to those who don't like liver, only a little more intense. I hated to waste the meat, so I tried about every recipe imaginable in hopes of eating it.

> "Some said I ate them; others thought I was kind of eccentric."

I found several interesting facts. First, I tried to blend the buck sheep meat with beef to stretch the beef a little further. The result of this experiment was the discovery that one teaspoon of sheep meat mixed with 10 pounds of hamburger rendered the total completely unpalatable.

One day, my grandmother gave me a recipe for sausage that looked like it might be the answer. Since a good friend had a slaughterhouse, walk-in smoker, blenders, and a grinder, we gave it a try. This proved very successful. Our sheep jerky sausage became extremely popular and was sold and given away for Christmas presents. The demand became overwhelming and we soon had orders for hundreds of pounds of "sherky." We decided to go into the sheep-sausage jerky business, fully convinced that we would not only become rich, but I would be able to obtain sheep horn for knife handles virtually free.

We then entered the real world. Going into production with food for human consumption involves more bureaucrats and rules than my corral has flies. The result? The great sherky business became a memory.

Some of the old bucks whose horns may one day comprise the handle of an Ed Fowler knife, such as on his multiple-quenched Pronghorn model, reside in Ed's corral.

How To Make Friends and Influence People, Sheep-Horn Style

The author uses his favorite handle material to repay an old debt

As the quality of my knives improves, my need for sheep horn for the handles increases. Word gets around that I'm interested in buying sheep horn. It doesn't take long for most everyone who raises sheep in my area to learn that I'm a ready market for dead and dying horned bucks. As a result, folks call whenever they have a dead or dying buck with good horns.

The wintertime phone calls aren't so bad but summertime calls are another story. They usually go along the lines of, "I have a set of

> **"When it comes to sheep horn, no hardship is too great."**

horns for you. One of my best old bucks died last night." I get directions and head out, not wanting to waste too much time, as dead bucks decompose quickly in the summer heat. Usually the callers are pretty accurate but sometimes they miss a little on pinpointing the day the buck actually died.

One time in particular, a 300-pound buck had been putrefying in the searing heat for about a week. His horns were exceptional, so I decided to load him in the pickup, flies, maggots, smell, and all. It was enough to make

The author's sheep horn knife and sheath rest on a choice piece of sheep horn.

If you own an Ed Fowler knife today, the horn handle may have come from among these old bucks on his Willow Bow Ranch.

your eyes water but, when it comes to sheep horn, no hardship is too great. We headed for home, the people driving behind us always staying well behind, those going the other direction always looking back at us due to the overwhelming smell of the dead buck.

Gas Wars

As we drove toward town, the buck reminded me of a run-in I'd had at one of the local grocery stores. I'd bought some canned sauce from one of them and when I opened the can, quite a bit of air pressure escaped, the gas not unlike that coming from the buck sheep in the back of my pickup. I'd gone to the manager and asked for a replacement can of sauce and he'd declined, saying it was impossible for one to go bad. I told him that I'd buy my groceries elsewhere and hadn't been back in the store for years.

I suddenly decided that it would be a good day to "make friends" with the manager. I parked my pickup, buck sheep in back, as close to the front door as possible, and walked into the store.

> "Pretty soon the store was virtually empty of customers."

It was kind of nice being back in the old establishment, air conditioning and all. There were a lot of people shopping. Strolling up and down the aisles, I found some things that looked kind of good, then decided I didn't need them and put them back on the shelf.

Pretty soon the store was virtually empty of customers and one of the checkers was walking around with some kind of spray can making the place smell like pine trees. The manager came up to me walking fast. I could see that he remembered the last time we spoke.

He asked if he could help me find what I wanted. I told him that I was looking for some of that "special sauce." He grabbed the front of the cart I was pushing, led me to the place where the sauce was, and tossed in a couple-dozen cans. I informed him that was more than I could afford; he said that it was on him, that I shouldn't even bother stopping at the checkout. Just get my pickup out of his lot!

Sometimes, sheep horn can make friends, and make folks more agreeable.

Knife Talk Philosophy

Dear New Knifemaker

An open letter to today's up-and-coming cutlers gets the rookies off to a good start

Recently, I received the kind of phone call I like to get. The caller wanted to learn how to make knives!

He'd held many jobs while working his way through school, so his job experience varied widely. He had a degree in art and education and was working on an advanced degree in psychology. He'd sold his artwork in many media and done well. He stated that his interest was in nature and her ways. Hunting with firearms, bow and camera, fishing, and backpacking were among his interests. I strongly encouraged him to join the world of knives, for his background and enthusiasm could well bring new life into the cutlery culture.

> **"There's joy in truly creating."**

I've talked to other newcomers to the world of knives and enthusiastically awaited their influence. Many times I find that they soon digress into making the same kinds of knives that are presently available. As a result, I've been giving some thought as to why this trend to the average is so prevalent.

The way I see it is, the first pitfall to the new maker is a loss in direction. This often happens when the newcomer first begins to make knives. His attention becomes focused on the technical aspects, such as steel selection, forging, grinding, and heat treating—regimens that are usually foreign to his way of doing things— and he becomes so focused on them that he loses sight of the tremendous gift of individual creativity that he could bring to the world of knives. He loses confidence in the ideals he once embraced and his knives become mired in the roots of tradition. His potential gift remains veiled.

The Initial Approach

As you develop the skills necessary to the basics of knifemaking, there will be times you feel that you know all you need to know when, in fact, it's only

When the newcomer first starts making knives, he may get lost in the technical aspects, such as forging, of knifemaking rather than focusing on his own creativity. Here the author forges a load shaft down to a flat bar.

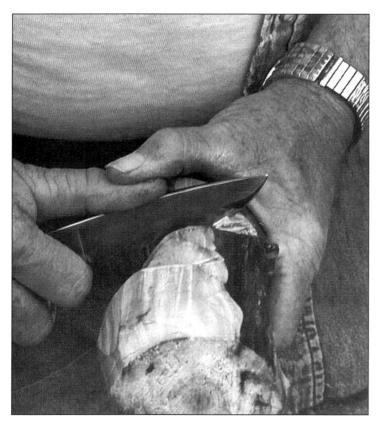
Know exactly how your customers use your knives, whether it's to whittle or whatever, so you can best know your customers and serve their needs.

"Build your kind of knife from your heart."

the beginning, a time to bring all your lifetime experiences into the knives you make and challenge yourself to do more. Through your research and total accumulated experience, you'll have the potential to make your own kind of knife. Should function be your goal, take the time to thoroughly test your knife doing the types of work with it you intend it to do. Constantly ask yourself "why?" and "what for?" when designing your knife. There should be no line wasted that will sacrifice purpose for the sake of design. There's no need to choke your personal Excalibur with unnecessary glitter. Your knife, when it comes from your heart, will speak for itself.

There's no requirement that you make every kind of blade—from dragonslaying swords to folding miniatures—in order to be a true master, nor is there any mandate to specialize. Look at them all. Study the knife in legend and personally question all you see. One day you'll know wherein your challenge lies. This direction may stem from a composite of all the knives you've seen, combined with your desire to achieve a certain functional quality or exceptional beauty.

The type of knife you choose to make matters not. Your "honest knife" will represent a sum total of all your experiences. I remember a friend who, while in college, became obsessed with the business end of all biting insects. He could be found with his eyes glued to a microscope at all hours, his drawings, and written comments carefully preserved in volumes of documents. I smile when I think of the sort of knife he would have made.

Build your kind of knife from your heart—it's your potential gift to the world of blades. Those who know knives, your style of knives, will come to you, while others will pass your table at knife shows as if it contains some kind of social disease. Don't be offended by their disdain; your kind of client will find you. When you copy simply to sell knives without feeling and knowing their true nature, you offer nothing new. Your status is second hand. Know your knives and you'll know honest self-respect. It's the joy of the struggle. There's equality and safety in stagnation, whereas there's joy in truly creating.

"Constantly ask yourself 'why?' and 'what for?'"

The sort of knife you choose to make will determine the kind of client you'll know. Should your specialty be hunting knives and you make affordable knives for hunting, you'll find yourself dealing with clients who are hunters. Woodcarving knives will bring woodcarvers to your table. The level of craftsmanship you put into your knives will also influence the nature of the client you serve.

Honest Knives

A knifemaker gives himself the kind of knife he wants to make. He'll own that blade from that point forward. When a client views the knife, he won't share the same thoughts that the maker knew during its creation. Honest knives are a personal expression. No two people can feel the same emotion when coming to know the same blade. The true

"Honest knives are a personal expression."

nature of appreciation is too personal to be known to more than one person. This doesn't mean that the art cannot be shared to mutual satisfaction. In making the kind of knife you want to make, you give your client a special satisfaction unavailable elsewhere. The knife will be good enough when it surpasses the expectations of your client.

It matters not what the majority of the world of knives thinks. I'm speaking of ego, your ego, the ability to stand alone if need be, to create something new from all the dreams of the past, then putting that special ingredient, the sum total of your personal experience, into something in which you believe and cherish.

Joe Szilaski, Hungarian by birth, learned many of the skills of the European craftsman. He worked as a "universal locksmith," the Hungarian equivalent of blacksmith, machinist, and tool-and-die maker. Among other things, he made knives and cleavers. The Hungarian Revolution forced his evacuation to Austria, where he worked in a quarry. (While American students were fine tuning pinball machines, Hungarian students were fine tuning their anvils!) A year later he came to America, where he continued to work with metals in a foundry producing metal sculptures. Joe has com-

"Your kind of client will find you."

bined European craftsmanship and the art of three-dimensional carving in all metals, tempered with the freedom of America, to bring a new expression of true art to the world of knives.

Should you, as a knifemaker, excel beyond your wildest dreams, it won't be because you worked hard or had the most elaborate shop. Your success will come from the fact that you're doing the very thing that you love doing the most.

Surely, the teachers in the world of knives are to share in the blame for failing to challenge their students to new frontiers. The goal of a teacher must be to teach with insight, driving his students through inspiration to seek higher levels without fear of failure, for everybody learns from failure as surely as they learn from success. This requires that the teacher provide all the information necessary to gain the basic skills—along with insight to the many questions unanswered—so that any knifemaker with the background and knowledge to bring new life to the world of knives will know that the opportunity exists, and he will accept the challenge.

How grateful I would be, as a teacher, if were I somehow able to inspire another knifemaker to elevate the craft higher than I can see.

What Exactly is "A Master?"

The author offers his assessment of what constitutes one

When I attended a forging tutorial taught by Bill Moran in Laramie, Wyo., years ago, I thought that I knew all there was to know about making knives. I was still an apprentice in the American Bladesmith Society and had to wait to test to qualify as a journeyman smith. Time passed and I received my journeyman smith stamp. At first I was very pleased, but there was still another waiting period and testing for my ABS master smith rating. When I received my master smith stamp, I once more thought that I "knew it all."

Still, somewhere in the back of my mind I knew there had to be more—but what? There were no more formal waiting periods, performance tests or judges that I would have to please in the future. I soon became acutely aware that there was a whole lot more information about knives of which I had either very little or absolutely no knowledge. The performance tests that I once considered extreme had become minimum standards.

As I pondered the issue, I was reminded of the parades honoring the Roman heroes when they returned from the battlefield. Behind each hero rode a slave who kept repeating a phrase indicating that all glory was fleeting. Why? The battle had been won; without a war to fight, warriors are without challenge and opportunity, so they must seek other fields of honor or else fade into obscurity.

What exactly is a master? I believe that, first of all, he is one who is truly alive in his work. If this

A maker who qualifies as a master to the author is Wade Colter. Wade's art bowie is 18 inches overall with a carved ebony handle and a damascus blade and guard. His address: Dept. BL11, POB 2340, Colstrip, MT 59323 (406) 748-4573. (PointSeven photo)

Few knifemakers can make the claim of master more legitimately than Blade Magazine *Cutlery Hall-Of-Famer George Herron, maker of this superb sub-hilt fighter. (Weyer photo)*

Should any of us think we know all there is to know about knives, we are as dead as the proverbial beaver hat, for we have lost vision. The best that we, any one of us, can hope is to master that one small corner of the world of knives that interests us, then explore that corner with all our energy and hope to come to know at least a part of our domain. The master must be aware of the relationship of knife to man in his time and place.

It must be remembered that no knife stands alone, for the knife is a tool and must be considered a companion of man in his time. In the 1800s, there were knives of America, Africa, Europe, Spain, and other regions existing all at once. Some were for warriors of actual battles, some for warriors cloistered to their civilized environment, some for armchair warriors dreaming to participate. Meanwhile, there were knives of the frontiers that faced their own challenges and more. The true master seeking to understand such knives must take time to keep the knives he loves in perspective as they relate to man.

A master must have what I would call a childlike enthusiasm for exploration. For example, consider a little kid on the frontier, wherever and whenever that was, who forged his first knife on the family's or neighbor's forge. For a time, the youth was a master of his knife, for he was completely obsessed with his task—and there is no joy as great as that first success to be realized.

The true master may blossom to the demands of the era in which he lives. Consider John Singleton Mosby, a hero of the Confederacy. He was a sickly youth, of slight physical stature who found, by virtue of the times, a cause that transcended his shortcomings. His accomplishments are legend and would have never been considered possible by those who knew him before the Civil War. The challenge of the period matched his hidden potential and a master was born.

What about the importance of intelligence? While intelligence is a fine attribute for a master, never underestimate the potential force of any

is true, I have known many masters in my life and read of many more who, while no longer on this plane, are still very much alive in the work and memories they leave behind. They all have had one thing in common: Each lived life to the fullest. In short, they were very much alive in the realm they chose to call home.

Mike Cleary ran a junkyard in my home town. It was a small mountain town, three blocks wide and three miles long, at the time hosting a population of about 1,700.

Mike positively loved his junkyard. Every piece of scrap iron had a history and a future. A visit to his junkyard was always an adventure. When I sought an old timber saw blade to render into knife blades, he never had a simple, plain, ordinary saw blade but one that had been used by "Paul Bunyan, Jim Bridger, Ten Day Jack" or some other legend, a blade that had been made from "the finest steel ever to cross the ocean on a three-masted trading vessel through the worst of storms." Mike was the master of Mike's junkyard and glad to be there. "Mike's Junkyard, Sweeter than the Hanging Gardens of Babylon," read one of his advertisements in the local weekly newspaper. Thus it was for Mike and this writer, also.

> "The true master may blossom to the demands of the time in which he lives."

> "It matters not whether others agree with him; what matters is that he makes them think."

Knowing One's Limitations

A master must be aware of his limitations. A master in the world of knives cannot be master of all the provinces in the world of knives, for those provinces are far too many and varied. Knives have been a companion of man since man began, and have been made from wood, bone, horn, seashell, teeth, stone, copper, bronze, many steels, and more.

Knife Talk Philosophy

individual who is absolutely dedicated to succeed. When one puts all his energy into focus, he is very well able to contribute greatly to the prevailing culture with a childlike gift of vision, bringing prominence into what may have been overlooked or ignored.

It is wrong to believe that masters are completely in command of their chosen specialty. They may be the best there is, but the true master is unfinished—and he knows it. The single most cherished aspect of a master is that he knows he has a lot to learn and he simply loves exploring the frontiers of his vision. It matters not whether others agree with him; what matters is that he makes them think. Most importantly, he dares to continually challenge himself.

Eric Hoffer wrote a book called The True Believers. True believers are masters and encourage all who will listen to pursue or join in their cause. Not all true believers are benevolent or correct, but they are obsessed with their cause and will spare no energy in the exploration of the frontiers they choose to command. True believers have led nations to prominence—and destruction.

Masters also fall from grace. For instance, Samson found his Delilah, David his Bathsheba. Essentially, masters are the same in all fields. They demand excellence and seek perfection, knowledgeably accepting legitimate compromises of reality while remaining loyal to their principles.

> "The true master is unfinished—and he knows it."

Obsession fuels their driving force to excellence. The true master is alive through his work rather than simply making a living.

The difference between a master of craftsmanship and a master of the true spirit of the knife lies in the nature and direction of the force that drives him. Supported by his knowledge that he is but a student earns one the title of master. As citizens of the world of knives, our frontiers are to recognize the artifacts of those who fit our individual tastes to qualify as true masters. Once those masters and their artifacts are identified, we can only enjoy the product of their journey in the frontiers of the knife.

When it comes to reproducing antique bowies, nobody has it mastered quite like Alex Daniels. His Broomhead & Thomas folding bowie repro features a mother-of-pearl handle, gold fittings, and engraving by Chris Meyer. Closed length: 4 1/2 inches. Daniels' address: Dept. BL11, 1416 County Rd. 415, Town Creek, AL 35672 (205) 685-0943. (PointSeven photo)

Mistakes I've Made

In the world of knives, humans interact with humans and, thankfully, perfection isn't absolutely required

The other day I renewed acquaintances with a lady who some years ago bought a knife from me for her husband. She said she paid the handsome sum of $75 for the knife and was glad that she bought it when she did because it and similar pieces have become so expensive. As I recall, the blade was stock removal, made from a used D-2 steel planer blade, and the guard was sculpted from an old bar of copper that I had picked up at a junkyard. The handle was, naturally enough, sheep horn. The knife wasn't bad for an early model but a long way from where I hope my knives are today.

On the way home I thought about a discussion I had that same day with a client who ordered a knife four years ago. I made the knife and my better half, Angie, sent the client a letter advising him that his knife was ready. He called me and stated that the piece was priced much higher than when he had ordered it. He said he felt that a deal was a deal, and he wanted to know why I was asking so much more money for the knife.

I explained that I'd learned a lot since the time he'd first ordered it, and that the knives I make today require more time, knowledge and, I hope, higher levels of craftsmanship than the knives that I'd built previously. He asked me what improvements I'd made and we became involved in a discussion that lasted about 30 minutes. Following our conversation, he said that he understood and agreed to take the knife.

His frustration with the price was my fault. When I started making knives, I had no waiting list and a lot of learning ahead of me. I didn't foresee the changes that would come to be in my knives, or the increase in time and labor that they would require. As my knifemaking knowledge and talent improved, I naturally believed—or hoped—that my clientele would know that my knives

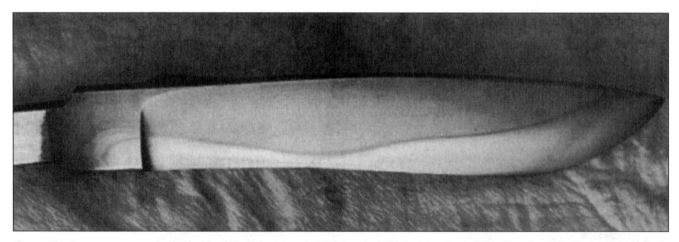

Once, Fowler gave away a knife that had been improperly heat treated. (Note how the temper line drops to the cutting edge in the center of the blade.) He believed there was no chance that the knife would ever be in circulation but it has resurfaced on the market not once but several times.

would increase in price as their value increased. Most of my clients know this is the case, but some don't understand without an explanation on my part.

I feel that, years ago, I should have made a statement that "these prices may increase as my talent, knowledge, and the quality in my knives increases." A statement to this effect could have helped avoid hard feelings on the part of some of my clients and sense of fault on my part.

It would be another story entirely if I were still making the same knife as I did when I started. I didn't enter knifemaking to sell knives; I entered it because I wanted to make them and hoped eventually to help achieve the Excalibur of my generation. The day eventually came when I had given my friends all the knives they wanted, and so I had to start selling some of my work.

Kids figure lots of things out and thus, as a young knifemaker, I found a way to subsidize my experiments and the development of my skills. When I started, I sold knives for as little as $5 and

> "His frustration with the price was my fault."

they may have been worth it at the time. That $5 for a single knife bought me another "Swedish timber saw blade," and I could make 20 more knives with enough material left over for a few of my early rudimentary experiments.

The good folks who bought my early knives provided me the opportunity to make more. Every knifemaker is beholden to his clients, for their support provides the financial and emotional nourishment that stimulates all makers to expand the frontiers of their craft.

The Knife That Wouldn't Go Away

Occasionally, all knifemakers make bad blades. A simple knife has little room for error but, as we makers push the performance qualities of our blades, the opportunity for error increases. Usually, I learn what I can from faulty blades and then destroy them.

However, one time I let one of my ranch hands have a knife that I used for an article concerning etched blades. The temper line obviously dropped to the cutting edge in the center of the blade. This por-

The author entertains his audience with a discussion of knives during a 1999 BLADE Show seminar.

Knife Talk Philosophy

tion of the cutting edge was too shallow to provide a high-performance knife. I photographed the blade and discarded it.

My ranch hand picked the blade out of the scrap pile and requested that I help him finish it as a knife for a Father's Day gift. I explained why it wasn't worth fooling with but he insisted. He had an old sheep horn that he had found for the handle. With steadily decreasing enthusiasm, I agreed to coach him through to a finished knife. I believed there was no chance that the knife would ever be in circulation, and that his dad would keep it forever. Meanwhile, I coached my ranch hand and the knife he finished looked a lot like one of mine.

Three years later, the knife surfaced on the market. Somehow, it ended up on a city street, was found, and brought to me for any historical information I could provide. I recognized the knife immediately and related the facts concerning it, and all was well for a while.

Several years later, I entered a sporting goods store and the knife was in a display counter labeled as a "Fowler knife." I enthusiastically explained the situation concerning the knife and the Fowler name came off the price tag. There were some hard feelings associated with the event. A year later, I again encountered the knife in another sporting goods store. This time my name wasn't associated with the piece, but I still received several phone calls from potential customers concerning its origin.

A few years later the knife again appeared, this time at the BLADE Show. Offered for sale as a "Fowler blade at a reduced price," it was still priced much higher than what it was worth. Several folks who were interested in my work asked about it. Finally, I agreed to subsidize the purchase of the knife if the man who was going to buy it swore to take the piece, use it to death, and never let anyone see it again. He agreed but somehow I still fear that some way that lousy, rotten blade will again come back to haunt me.

Haunting Scenario

I've seen the same scenario haunt other knifemakers. One time I was at a gun show and observed a damascus knife for sale identified as coming from the forge of a good friend of mine. It had all the earmarks of one of his knives, but something was wrong—the blade didn't look right. I asked to see the knife and found it covered with a greasy substance. I attempted to wipe the blade off and a loose weld on the side of the blade stuck in my finger. I knew my friend would never finish a blade of such poor quality. I photographed the knife and sent him a picture of it. He recognized the blade immediately and related the now familiar story. He had made the blade, knew it was defective, and had absolutely no intention of finishing it into a knife. A good friend of his saw the knife, asked to keep it as a memento of their friendship, and my maker friend gave it to him, never dreaming it would be put into circulation. Somehow, someone made a handle for it, strongly resembling my friend's style, and sold it as a knife made by my friend.

The Willow Bow Ranch Knifemaker's Absolute Rule: The maker should never, ever, no matter who begs or what circumstances come to pass, let a knife blade out of his/her shop that he/she knows is defective. There is one viable alternative: The maker should mark the blade in an eternal and obvious manner "experimental" or "defective gift to friend," or whatever will absolutely identify the blade as an exception to what the maker offers for sale.

These aren't the only mistakes I've made; there have been many more, and new mistakes await me. In the world of knives, humans interact with humans and, thankfully, perfection isn't absolutely required.

When First Place Isn't Necessarily the Best Place: Part I

In Part I, the author critiques knife show judging competitions

"Throughout the whole history of art, committees and juries, whoever composed them, have failed to pick winners."—Robert Henri, The Art Spirit, *Westview Press*

There are two types of competitive events that can be a part of many knife shows. First and most prevalent are the judging competitions where knifemakers make and enter pieces that fall under various classes, such as bowies, fighters, utility knives, folders, etc. These events, judged by individuals whom the sponsors of the show feel are qualified to evaluate blades based upon the individuals' experience in the world of knives, are most popular. The other event type consists of competitions where knives are judged solely on performance. The most significant example would be the rope-cutting competition.

In order for all these events to remain in perspective, I feel that some issues need to be considered concerning knife show competition in general.

It has been a long time since I attended my first knife show as an exhibitor. I was like a kid at his first county fair, in awe of the diverse nature of the people and events that evolve at any gathering.

One exhibitor had not sold a knife and very few people had shown much interest in his product. He had entered the competition for the best new maker. When the winners were announced, he was judged the winner of his class. Immediately, his table was surrounded by clients with new-found enthusiasm for his knives. Obviously, the judges' decision influenced some clients' opinions of the maker's knives. Some enthusiasts of the cutlery industry rely on such competitions to influence their purchases, and winning the competition may be good for the victor's business.

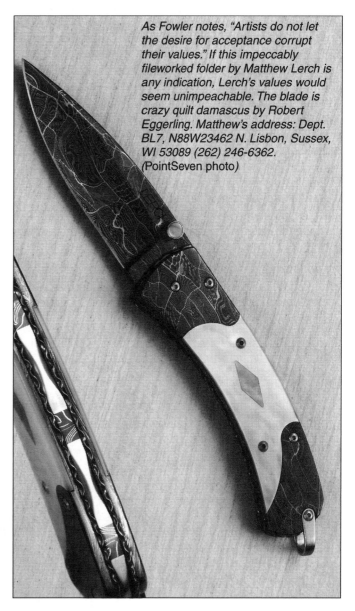

As Fowler notes, "Artists do not let the desire for acceptance corrupt their values." If this impeccably fileworked folder by Matthew Lerch is any indication, Lerch's values would seem unimpeachable. The blade is crazy quilt damascus by Robert Eggerling. Matthew's address: Dept. BL7, N88W23462 N. Lisbon, Sussex, WI 53089 (262) 246-6362. (PointSeven photo)

I have entered the judging competition at two knife shows. I won one event and a trophy hangs on my wall. At the second show, I do not remember what class I entered but I was not the winner. The judges presented the makers who entered the competition with written comments concerning the makers' knives. I read the comments many times for several weeks, considering the judges' suggestions. Finally, I reasoned that, should I decide to follow the direction the judges' comments indicated my knives should take, I would not be making knives of my choosing but of the judges' design. I did not opt to be a rebel, only to be true to the knives of my dreams.

Not too long ago I was asked to judge at a show. I thought about it for some time and then, after much deliberation, decided to accept the invitation. As it turned out, I was not selected to judge. Still, it had been rumored that I might be one of the judges. While at the show, one of the makers who entered the competition commented to me that he had made a knife that he knew would be to my liking. At that point, I came to appreciate the fact that I was not selected to be a judge, for I did not offer to be a judge with the intent to influence the makers in their style, but to offer my opinion concerning the design of the knives they chose to make.

> "Awards all too often demand that the artist please the judges rather than himself."

These three events, many conversations, and a lot of time thinking about knife show competitions have prompted me to consider the competitions as a topic of discussion. I fully realize that many excellent makers enter knife competitions and find them highly rewarding and exciting. Many times the knives that earn awards are tributes to the cutlery industry and the trophies are earned by hard work and dedication. I congratulate the winners and admire their commitment. On the other hand, I feel there are some aspects of this type of competition that should be open for discussion.

Unforeseen Consequences

How often have the judges been accurate in picking long-term winners when selecting the best new maker, for instance? While there is no documented record of which I am aware, I feel that all too many of those chosen best new maker have dropped from the industry long before those who also competed or did not compete at all. Could it be that the fledgling maker loses some of his appeal as a creative artist when he makes knives to the satisfaction of the established authorities? I feel that if a new maker is deserving of recognition, it must be because he has his own ideas.

To many, knives that finish right behind the winner—as this Bill Herndon Persian fighter did in its category at the 1999 BLADE Show West—can be just as valuable or more so than those that win. Herndon's address: Dept. BL7, 32520 Michigan, Acton, CA 93510 (661) 269-5860. (Weyer photo)

One also must consider the influence the judging has on the makers who do not win, for the direction of their potential contribution surely must be influenced as well. Could it be that judges and awarding prizes actually result in a form of control concerning the direction new makers' contributions to the industry will take, and that this control actually may be harmful to the career of the artist, as well as adversely affect the long-term direction of the industry?

Awards all too often demand that the artist please the judges rather than please himself. Creativity is an individual and private affair and is most viable when it comes honestly from the heart of the artist. One must always remember that artists are not in competition with each other but with themselves.

I feel that many new makers have opinions and dreams of their own design that belong to them exclusively. It is their individual creativity that will not only keep them inspired, but also breathe longevity and passion into their work and the establishment of the knife as true art. Each maker should exercise the freedom and know the satisfaction that comes to those who make their own kind of knife, a creation that pleases him without the influence and prejudices of another generation. Should we, as supporters of the cutlery industry, wish to encourage the young artists of the world of knives, ours is not to judge but to encourage by becoming genuinely interested in their efforts and willing to accept their personal and individual contributions.

When the established leaders of the world of knives seek to judge the new breed, the utmost care must be exercised to avoid perpetuating existing prejudices in the art.

In all probability, official competition has been a companion of art since man first developed the ability to think abstractly and express his desire to describe what he saw and felt with his hands, or voice, or in his dance. The problem with art competition is that there often is no yardstick, no scale and no scoreboard to adequately measure the message called art. Quality is in the mind and talent of the creator, and its acceptance in the eye or ear of the beholder. Artists seek recognition in order to share their vision and many, thanks to their dedication, do not let the desire for acceptance corrupt their values. When creativity and acceptance join in complete honesty, one can ask for no more.

America is becoming an art country, and a great many Americans wish to participate in art. They come from all walks of life and have varied interests. For many, judging is a significant part of their participation in this art movement. Many knifemakers compete in the judged events, and many more choose not to compete. When a maker decides to compete, he must realize that he is placing his talent in the judges' arena and, if he is to win, he must "play the game." It is impossible to please all judges with the kinds of knives the artist chooses to make. Rejection is a part of being an artist.

As a knifemaker, if you enter such a contest and do not win, do not be discouraged or hold a grudge. Smile, be the critic's friend, have fun with it, and the competition will remain in perspective. Knives as art are simply the maker's way of sharing life with others. Do not dwell on rejection, listen carefully to the critic, consider the message, then make your kind of knife and enjoy it.

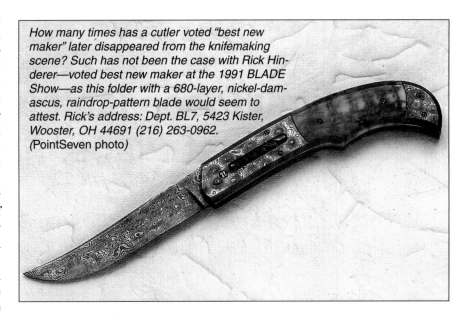

How many times has a cutler voted "best new maker" later disappeared from the knifemaking scene? Such has not been the case with Rick Hinderer—voted best new maker at the 1991 BLADE Show—as this folder with a 680-layer, nickel-damascus, raindrop-pattern blade would seem to attest. Rick's address: Dept. BL7, 5423 Kister, Wooster, OH 44691 (216) 263-0962. (PointSeven photo)

> "There is no yardstick to adequately measure the message called art."

Knife Talk Philosophy

When First Place Isn't Necessarily the Best Place: Part II

The world of knives needs those willing to pit their originality against the fashion of the day

Editor's note: In the final installment of his story, the author critiques knife-performance events and revisits the influence judging competitions have on individual knifemaker creativity.

Of great interest to me are the knife-cutting or performance competitions. This is one arena of knife competition where empirical evaluation is possible and true winners reign. Some of the competitive events are realistic, while others leave me asking questions such as "why?" and "what's the point?" To keep the competition knives in perspective, it must be remembered that the knives are specialists—they have to be, for these are some highly contested events.

For example, when it comes to the hanging-rope-cutting competition, the knives must be designed to sever the rope with extreme efficiency. To consider a winning knife in this competition to be anything other than a finely tuned specialist is misleading. The best rope cutter cannot be an adequate camp knife because the best rope-slicing blade requires a hard, thin edge, while the stout camp knife must have a thicker blade that is tough and strong for chopping and other camp chores. A realistic analogy would be to compare a Kentucky Derby winner to a ranch horse.

These specialized high-performance knives speak highly of the ability of the individual knife-

Ed Schempp's knife excelled at cutting rope—Ed won the rope-cutting competition with it at the 1999 Oregon Knife Show. However, when the edge was used as a camp knife, a job for which the edge was not designed, it chipped. This was not Schempp's fault. It was instead a case of using the wrong tool—a rope-cutting specialist—for the wrong job—chopping. (Gallagher photo)

80 Knife Talk Philosophy

As the author notes, each knifemaker should realistically evaluate his or her knives, be it through function or history. Mark McCoun's damascus natural folder features mosaic damascus bolsters and a contrasting damascus blade. The damascus is by Robert Eggerling. (PointSeven photo)

maker to create a fine-tuned hanging-rope-cutting instrument. The winners of these events well know the intricacies of edge geometry, functional balance, and heat treating.

At the 1999 Oregon Knife Show, Ed Schempp competed against the cream of the crop when it comes to knifemakers who know what a rope-cutting knife has to be. With his knife, he worked his way from one length of 1-inch rope to eight lengths, cutting a total of 36 lengths of 1-inch rope without sharpening the blade. The cutting edge had to be hard, the edge geometry perfectly and specifically designed for the task. Schempp's outstanding knife did the job it was made to do. When this magnificent lady was put to the task of chopping seasoned wood, the edge chipped. Was the knife a failure? Most absolutely and certainly not! She was an outstanding success at doing the job she was designed to do.

Ed created a fine-tuned specialist, a tribute to the bladesmith's art. His knife did not fail to perform the task it was designed to achieve. It was a rope-cutting specialist, not a camp knife. Schempp well knows what it takes to make a knife that is up to any task, from chopping bone to slicing tissue paper, and proved his ability in the competition.

To properly judge knives, they must be considered in perspective. Know them for what they are meant to be.

Spirit of the Knife

An understanding of the spirit or true nature of the knife is necessary to keep the blades of history in perspective. There have been many knife fads. The 1800s knew the bowie, the 1950s the switchblade, and then there were the survival knives of the 1980s. Today, tactical folders know the spotlight. These pieces must be considered for what they were or are.

> **"Do not judge your knives solely by the quality you see in the work of others."**

To consider, for example, the Sheffield bowie a knife of the frontier is erroneous. The Sheffield bowies were instead mass-produced symbols of the frontier and sold to men who wished to share a part of that heritage. Such bowies are studies of elegance and grace, fit and finish, high levels of craftsmanship, and were props for the photographer. They were knives of civilization but not truly knives of the frontier, for most of them lacked the most critical aspects of absolute dedication to function and dependability.

Some years ago I owned a fine Sheffield bowie, probably dating from the mid-1800s. Not knowing the value of the knife, my father gave it to a youth for helping him clean up his yard. That fall the youth brought the knife to me, wanting me to repair it. The edge was badly chipped. I asked him what he did with the knife and he replied that all he tried to do was dress out a mule deer. I attempted to flex the edge and it chipped instantly, the edge being much too hard for the easiest field use. The knife was from one of the well-known companies and may or may not be representative of all Sheffield bowies. Still, judging by the design of many such knives, they were not intended for serious use on the frontier. It is OK to revere them but we, as knife enthusiasts, do a disservice to history and the knives themselves when we fail to preserve their memory for what they truly were.

Some knifemakers, such as Ray Appleton, John Nelson Cooper, Bob Loveless, Frank Richtig, Rudy Ruana, and William Scagel, hold or held true to their personal voyages of discovery and make or made their kind of knife free from the shackles of conformity and tradition. Knives from their hands are as exclusive as their signatures. Their truly unique knives have escalated in value far beyond the knives of tradition. Such makers are or were dedicated to originality more than to the fashion of the day. The fact that these men bring or brought new life into the world of knives is a tribute to

Jerry Lairson Sr. won the award for best fighter at the 2000 Arkansas Custom Knife Show with this stag stalwart. The 8 1/2-inch blade is damascus and the guard is wrought iron. (Ward photo)

them. The world of knives needs men who are willing to pit their originality against the traditions of the day, for this is the basis of growth and a viable future for everyone in the cutlery community.

I would caution the new knifemaker: Do not judge your knives solely by the qualities you see in the work of others. Just because some makers have been around a long time does not necessarily make them authorities. I am at times in awe that many have become prominent in the knife culture in spite of their proclivity to consistently ignore knowledge, as well as their potential individual creativity.

Should the maker wish to devote his talents to one artifact of any culture, he needs to study carefully the knives of his choice. If you are a knifemaker, do not limit your study to the knives only. Consider the people who used them and their culture, for the partnership with man is the most significant aspect of the knife. The discovery of the nature of the partnership with people will unlock the door to memories, that something special all knife enthusiasts wish to share. As you come to know the knives of your choice, you will find there is boundless room for creative expression.

The techniques of making a knife extend from absolute simplicity to the extreme limits of technology. Embellishment can be achieved very effectively with no more than a piece of sandpaper to the most expensive stones and precious metals. Once you find your direction, realistically evaluate your knives, be it through function or history, thoroughly test your knives according to your goals and enjoy the voyage. Do your homework and the influence of judges and competition will remain in perspective, for you will know more about your knives than any judge. Your expression of freedom, that special part of you, will impart honest spirit in your knives. It is the spirit of the honest blade that will add to the cutlery culture, attract more citizens to the world of knives and ensure its future. Freedom is the essential element of creativity. Use that freedom and enjoy life in the world of knives to the fullest.

Beware of the Medicine Man

Sharing knife knowledge bodes well for all

The brightly colored medicine wagon rolled down Main Street, dust rising from its wheels, as the townsfolk gathered for the arrival of the Pony Express and news from the rest of the world. The medicine man climbed up and stood on the back of the wagon and began his pitch.

There are few more noble goals in the world of knives than sharing knowledge, as Wayne Goddard, left, does while instructing at an Alabama Council Symposium.

He offered, for the mere price of a week's wages, a new "secret potion" that was a 2,000-year-old Chinese remedy. It would cure cancer, syphilis, rheumatism, impotence, and attract women like flies, he claimed. An articulate speaker, he attracted the townsfolk and, for the afflicted, inspired new hope.

As the people wasted their hard-earned money buying bottle after bottle, the old town doctor stood at the back of the crowd. He sadly shook his head while he watched them, many patients of his, lay out cash money for the worthless cure, the same people who hadn't had the means to pay what they had owed him for years. With mixed feelings he watched as they chased hopeless dreams. With great desire he wished some such cure existed, as it would have saved the lives of many of his former patients whose company he greatly missed.

The town marshal stood next to him as the event transpired. The aged doctor spoke of the travesty that they were witnessing, suggesting that the charlatan should be run out of town. He noticed that the marshal was rubbing his shoulder, a shoulder that had given the lawman some pain since the big bar fight years ago. The look in the marshal's eyes told the doctor that his suggestion fell on deaf ears, ears which would soon turn red with the after effects of the secret potion.

While you might smile when you hear of such events, the cancer that supports the charlatan still eats on mankind today. This seemingly trivial event is more malignant than it appears on the surface, for by its very nature the survival of man is threatened. He who sells his product on the basis of "secrets" either sells false dreams or is the keeper of a secret that, should it have any real benefit, robs all men of better times solely for his own gain.

The medicine man illusion survives in all worlds, and the world of knives is no exception.

I could very easily take a torch, play it on a blade, and create a temper line on the surface of a specialized steel and claim a new "secret" process that few could duplicate, and make great claims of its quality. The knives would sell without a doubt but my method would only add a little smoke, impairing everyone's vision of a better blade, for should I share my knowledge, another maker could very easily advance my discovery to even greater heights and all would benefit. Whenever knowledge is shared, it's also gained, as information shared grows beyond the vision of any single individual. The danger of competition emphasizing secrets or exclusive materials lies in the fact that it takes most from those who can ill afford the loss and preys on a great weakness of man, greed-humanity's greatest threat to survival.

That people depend upon each other, even in these "modern times," is revealed when one transformer can cause an electrical outage involving communities hundreds of miles away. Should one man's "secret steel" or knowledge gained from his "secrets" lead to the prevention of hardship, it's paramount to all men that it be shared.

Ultimate Survival

One day, man as a species may face the ultimate test of survival, either due to the whims of nature or man himself. Should this occur, you can bet that the "medicine man" will be out for his own interests and manifest little compassion for those who fall prey to his illusion. Mankind will know no comfort from him. Man's only chance for survival lays in the ability to develop an altruistic sense of community wherein humanity, love, and quality survival for all is the goal.

The world of knives is but a small pond in the immense waters of life. Still, leading by example in the world of knives can be a road map for others to follow. Should knife enthusiasts of all stripes make a difference in but a small percentage of the rest of man, they will have contributed well.

Do I have secrets? Yes, thousands of them. Every time I make a knife, a new finite method makes itself known to me. I don't keep these secrets willingly. Only by their nature do they not become known. They're the secrets that come to mind during the thousands of solitary hours making knives—some of consequence, many insignificant when considered alone. "A little more pressure here, blend the edge to tip this way, a change in sequence, 'a different' hold or stroke with a hammer," the list goes on and the mode of discovery is well known to all knifemakers. I would offer all these "secrets" willingly to those who are interested, but for want of words and time cannot find a means of sharing them. Were I to write of each and every one, most readers certainly would be bored. Still, should anyone ask, he won't be denied. Fortunately, many knifemakers share this philosophy.

All blade enthusiasts swim the same waters. Those who buy the knives of the "medicine man" surely won't be interested in my blades. Nor do I seek to be associated with them, for they support the ideals for which I have nothing but disdain. All knifemakers compete by virtue of craftsmanship, design, and desire. My greatest competition is with myself.

When you share time and knowledge honestly with those who share your vision, you know good times with friends—and, I hope, influence the future nature of man toward a sense of community. The future will be shared by my friends, the coyote, and the rattler, proven survivors, as will the breed of man who's developed the skills of altruism and the hunter to carry on the species, Homo Sapiens, hopefully to a more perfect civilization.

> "The medicine man illusion survives in all worlds."

Guilty Until Proven Innocent

*Another anti-knife school rule punishes
the good, not the bad*

Joe was 13 when he entered his seventh school this past year. He has had it tough, moving from foster home to foster home most of his life.

Joe cannot compete scholastically with other kids and has been placed in what is termed "special education." While Joe will, in all probability, never know the glory that the public education system calls scholastic success, he does have some talents. He is gifted when it comes to mechanical aptitude, he likes to work his hands, seeks friends, and according to family members, has never displayed hostility toward others.

Joe entered his new school in October 1998, missing the first seven weeks of the school year. As a result, he did not receive the carefully drafted documents concerning school rules.

Joe's new foster mother has raised many children, both her own and those of others. One of her children learned to make knives years ago in the

Fowler said that if children are taught by the kinds of people who do not know the difference between a weapon and a tool, America will know the same kind of failure that has condemned other great cultures. Here is Paul Crofts whittling, accompanied by his dog.

same school that Joe had just entered. Joe's new foster brother began teaching Joe how to make knives. Joe took to the craft enthusiastically. His foster mother has a kitchen full of knives that Joe's foster brother made, and Joe saw making knives as a chance to become a success, a very important achievement for a boy in his position.

He worked hard on his knife and soon it was ready for a handle. His foster brother saved him some deer antler and Joe was anxious to put his first handle on his first knife.

He lives one mile from the school and works at a lumberyard after school. He left home one morning this past March, his project in his backpack, with the intention of going straight from school to the lumberyard to work on his first knife. This was convenient because if he had not taken the knife to school, he would have had to walk one mile home, then retrace the same mile back past the school to the lumberyard to work on his project.

Joe knew nothing but enthusiasm. He was a success and had mentioned his project to two of his new school friends. When he arrived at school, he proudly showed one of his friends what he had made. As far as Joe knew, he had nothing to hide. He was not sneaking the knife into school, he was sharing his success with his friends.

School started, and Joe put the knife in his backpack and the backpack in his locker. He did not know he was breaking any rules. He had never been told that knives were not allowed in school, nor did he know knives to be anything other than a tool, a craft project, and a chance for knowing the joy of making something with his hands. To him, it was art that has been a part of mankind for as long as man has had thumbs.

Another student, who watched as Joe showed his knife to his friends, told a teacher. The teacher told the principal and Joe's locker was searched. The knife was confiscated and Joe was taken from class and immediately suspended for violating "federal rules" concerning knives in school.

Joe's enthusiastic success instantly became a failure. It is a failure that he cannot understand because to him the knife is a tool used at the family table in the kitchen. What is more, the knife he made was an accomplishment, a success that those who know knives or have made something with their hands can fully appreciate. Joe cannot play the piano, write brilliant poetry or even understand what kids his age learn easily. If he is going to make his mark in this world, he will have to do it within the bounds of the skills he has. As it stands, he is branded an outcast for doing nothing more than being a boy with special talents.

The Wrong Lesson

By enforcing the rules that even kids know make no sense, we as a people are teaching our future generations the wrong lesson. Is it any wonder why some school age children are angry? Is it any wonder that they become couch ornaments in front of the TV, or drift to drugs to substitute for the thrill of realistic accomplishment?

Do not get me wrong; I have no sympathy for punks who use a knife as an offensive weapon against another person. I firmly believe that human predators who choose to attack an innocent victim should immediately know the full force of the criminal justice system.

Arbitrary rules drafted thousands of miles from our shores by bureaucrats that gave no heed to the needs or desires of the inhabitants of the new colonies sparked a revolution, and thus began the great experiment known as America, land of the free! Where did the rule concerning knives in school originate? Was it put to a vote? Did the electorate even know it had been enacted? I approached the local school administration officials on the subject. They could not produce copies of the "federal rules" they claimed to enforce, and referred me to the local library, where my quest for documentation proved fruitless.

I contacted the Rutherford Institute, which has championed the cause of children victimized by what has become known as zero-tolerance rules of safety in the schools. I learned that schools are safer today than they were in the 1980s.

As for the "federal rules" cited by the local school officials, my impression was that the rules were conceived in 1994 or 1996 in what is referred to as the "Gun-Free Schools Act." These federal guidelines mandated that the schools use discretion in their enforcement. "Discretion" entails the right of all citizens to due process, that the intent of the students' actions be considered before judgment, and that remedies must be appropriate to the students' acts. Somehow, it seems to me, that some school administrations have lost sight of common sense when it comes to discipline.

I feel that if the children are taught by the kinds of people who do not know the difference between a weapon and a tool, America will know the same kind of failure that has condemned many former great cultures throughout history.

While concerned, I feel that if the parents of children in Public School 63 in another city decide to ban knives in their school, that is their prerogative, for it is essential that communities maintain control of their schools so the freedom that conceived America and the creativity that makes her great can nurture her future. It disturbs me greatly when some bureaucrat who knows nothing of the culture here in Wyoming, or your children's school, imposes a rule on the schools of an entire nation. The bureaucrats, the citizens' paid employees, have become dictators. If we as citizens do not draw the line and stand for our independence and create rules based in reality and justice tempered with understanding, I fear we will surely lose our future leaders—the children—to carry on the great American experiment.

Joe does not stand alone. There are many Joes in many schools who have been subjected to the same malignant milestones of injustice as the American culture treads a downward path led by bureaucrats dictating rules that are not solidly founded on reality. Any tool can be used as a formidable weapon. To forbid its presence among all people is legislative folly. Tools are needed to prepare, build, fix, and create. Tools make life more convenient, and the knife as a tool can and does save lives regularly—problem is, no one keeps score when a knife saves a life.

I strongly believe that, as parents, we are responsible for our children's education. As citizens, we must all take an active interest in what our schools are teaching and what is lacking therein. Whenever injustice takes place in our communities, we must be as outraged as if we were the primary victim of that injustice, or soon there will be no one left to protest. Write letters to the local newspaper, attend your school board meetings and voice your concerns, support the creative teachers of your schools, call your legislators. Let them all know that you are concerned, and vote!

> "Some school administrations have lost sight of common sense when it comes to discipline."

The Downside of Mint Condition Blades

You can truly know a knife only by using it as it was meant to be used

My article about the "Lewis and Clark knife" in the June 2000 BLADE® received more reader response than any article I ever have written. I wrote it more than a year ago and sent it to the editor questioning seriously whether it should be published or not. My fears were that it would aggravate BLADE readers or be inappropriate somehow. As it turns out, many appreciated it and had a lot of fun sharing the emotions it obviously generated with their friends. Some commented that, knowing me and my obsession with function, they felt that if I had actually possessed a knife from the Lewis and Clark expedition, I very easily may have explored its performance potential as depicted in the article.

Recently, Gordon Minnis—author of an outstanding book for any knife enthusiast, American Primitive Knives 1770–1870—and I were discussing the issues that surround such a treasure. He asked what I would have done if the knife had been authentic. I have pondered some time over his question.

I believe that if I did in fact come in contact with such an artifact, I would first thoroughly explore and photograph every aspect of it. I would definitely sharpen the knife and cut some rope with it. Then I would learn as much as possible about the nature of the knife by having Rex Walter run a chemistry on the blade to discover the composition of the steel, as such a test would provide some indication of authenticity. I also would request that he perform a Rockwell hardness test to determine the hardness of the blade. The tests would not damage the knife significantly and would bring to light some valuable technical and historical information before the piece could be declared a national treasure, and be committed to some institution for posterity.

The Swenson Conversion

Recently I experienced a similar situation involving a handgun. An elderly gentleman decided to sell his extensive firearm collection. Most of the firearms he had sold up to that point—and ostensibly continues to sell—were made between 1850 and 1980. His appreciation for excellence is obviously well manifested in the quality of the firearms he collected.

Not long ago, he brought in a Swenson conversion on a Series 70 Colt .45 ACP automatic pistol. As many readers may not be familiar with the legend of the late Amend Swenson, I will state that he was one of the top U.S. gunsmiths who devoted his talents to improving the accuracy of the Colt 1911 automatic pistol while still maintaining the gun's 100 percent functional reliability. He invented the ambidextrous safety and was a truly fine gentleman devoted to performance during the mid-1900s.

Anyhow, the pistol in question was in the same condition it had been in when it left Swenson's shop. The owner had simply oiled it well

Thanks to Dr. Jim Lucie, the author once was able to sharpen and cut with a premium William Scagel knife—though not this one—and thus learn of its true nature. This Scagel knife was Scagel's personal hunting piece. It has a silver-mounted guard and silver cap on the end of the tang. (photo courtesy Dr. Jim Lucie)

and put it away, never having fired it. The pistol's high quality of workmanship was obvious. At one time I shot many targets in competition and yearned for a handgun of such quality. The owner's asking price was $720, which to me was then—and still is—a lot of money.

Visions of the pistol remained prominent in my thoughts for the next few weeks. I kept hoping that someone would buy her and remove the temptation, but every time I walked into the store she was there to greet me. Finally, the desire to explore her qualities outdistanced my fear of a questioning wife, so I bought the pistol. (Mrs. Ed [Angie] Fowler's note: I object to the adjective questioning on the grounds that I only question Ed's gun purchases when we have outstanding bills that must be paid!)

As I was getting ready for the 2000 BLADE Show, I did not have time to shoot the pistol but I did take her apart several times, cleaned and oiled her, and devoted my break times examining her workmanship. Then I found myself caught up in the dilemma of keeping her in pristine condition or actually shooting a mint condition firearm that can never be replaced. The day before leaving for the BLADE Show, I decided she was made to perform and ready for the opportunity for which she longed.

I walked out to my private shooting range with the Swenson Conversion and 250 rounds of my favorite hand loads. I fired the first shot, picked up my binoculars and was delighted to see a bullet hole on the paper target, 7 o'clock in the white, just out of the black bull's-eye. I then fired a group that was small enough to make any target shooter smile. I was immediately impressed by her smooth functional qualities. A few sight changes and I started putting the handgun through her paces. Bill, our hired hand, walked over, smiled, and commented that it looked like I finally had decided to shoot the pistol. He picked up the binoculars and spotted my shots. His comment: "You are eating the bull's-eye out of the target." My spirit soared with the eagles!

It did not take long before I was out of ammunition. As I walked back to my shop, I suddenly felt sorry for the man who previously had owned the pistol. Though he had possessed it for more than 25 years, he had never known the privilege, the absolute joy of shooting, and knowing the performance qualities of one of the finest handguns I have ever known. Also on my mind was the fact that Mr. Swenson knew what he was doing when it came to making fine target 45s. I feel he would have been proud to know that the performance qualities of his workmanship were appreciated and had met the expectations of a shooter he would never know. Through his workmanship and my enthusiasm we had shared some quality time together.

Use It or It's Abused

The same considerations apply to any artifact of history, be it a tool or a document. Tools are meant to be used, and books need to be read and their thoughts shared. Such items only have value through their service to man. I put as much time into the basic nature of the steel that goes into the knives I make as I do into their cosmetic attributes. High performance is the goal that drives me. Nothing pleases me more than to hear that one of my knives has met the performance expectations of its owner. Many a knife is pleasing to the eye and feels kind to the hand that holds her. To add the quality of performance to the knife puts the package together.

I well remember the time, thanks to Dr. Jim Lucie, that I was able to sharpen and cut with a premium William Scagel knife. By being permitted to enjoy the true spirit of the Scagel, I was able to fully appreciate the depth to which the maker dedicated his finely developed knowledge and craftsmanship to bring an honest knife to my hand.

There is much interest in "mint condition knives." Their value is always greater when they have been well cared for and left unused. Still, I cannot help but feel that those who fail to use their knives are missing out on the greatest pleasures that our favorite tool, the knife, has to share with man—service and memories.

There are a lot of knives out there that were conceived, designed and intended only as art, and "cut" was never a part of their heritage. Such knives are best when admired visually, for that was their purpose. When it comes to knives that were designed by men who explored the frontiers of function, we can only truly come to know them through using the knives as they were meant to be used.

Brothers in Steel

In the final analysis, all knives are part of one big, sharp family

Some seek to forecast the future of the world of knives by basing their predictions on the popular pieces of the present. Many knives that are highly sought after now will become insignificant tomorrow, while some little known today will become very valuable and highly sought after.

For example, I recently noted a handmade knife in Northwest Knives and Collectibles' April catalog. Owned and operated by Bill Claussen, Northwest Knives and Collectibles is an outstanding cutlery store in Salem, Ore., offering fine blades, both old and new. In describing the knife, the catalog spoke of "absolute mastery of the bowie." Made more than 150 years ago by a J.R. Rogers, the knife was valued at $4,995 in the catalog. I'd never heard of the maker but, based on the workmanship and elements of design manifested in the bowie, I felt he had surely left his mark in the world of knives. I wanted the opportunity to get to know the knife and maker better.

I was a little bashful about sharing my find with my wife Angie and softened the news by talking about the BLADE® article I planned to write about the knife—along with the $1,500 edged treasure she'd just purchased for me to be used for several articles. Fortunately, she thought I was kidding and dismissed the discussion. As the "new" old knife was only going to cost a little over $4,000 more than I had paid for my old pickup, it didn't seem like such a bad deal.

Whether they use carbon steel or stainless, all makers are brothers in knives. Ed Baumgardner offers a 154CM hunter in the author's favorite handle material—sheep horn. Overall length: 7.5 inches. The engraving is by Mike Branham. The maker's address: 128 E. Main, Dept. BL9, Glendale, KY 42740 (502) 435-2675. (PointSeven photo)

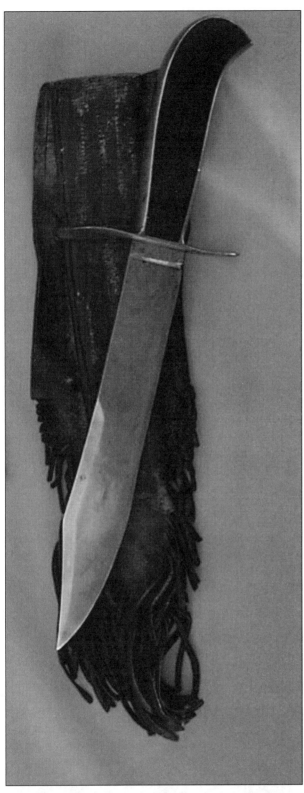

This J.R. Rogers knife was offered for sale in the Northwest Knives and Collectibles April catalog for $4,995. The stamp on the blade reads "J.R. ROGERS MAKER BUTTE MONTANA TOOTHPICK." Rogers lived in Butte in the 19th century. The knife is 13 inches overall. (Northwest Knives and Collectibles photo)

"Truth builds confidence, friends and repeat customers."

Luckily, fate intervened and another man spoke for the knife. Actually, several gentlemen spoke who knew better than I what she was worth. The last estimated value I heard on the bowie was several times the initial cost, and the increase was realized in less than two weeks. Those who sought her are serious students and collectors of ladies of her lineage. Her true value wouldn't have mattered because, if I had liked the knife, she would have never been sold—at least not by me—and I would have a $5,000 investment Angie and I couldn't afford at the time.

The knife in question is but an example of a long-term value that's firmly established in the world of knives. Superior workmanship, artistic quality, and sound design are usually good investments. At the time the knife was made, between 1840 and 1850, she was in her prime; newspapers and dime novels were establishing her and her kind in legend and history. Her popularity (value) would be greatly enhanced by publications such as *The Antique Bowie Knife Book* and such movies as *The Iron Mistress*. Meanwhile, the maker may have had difficulty making ends meet at the time he fashioned the knife. Today, his work is highly sought after and well deserving of the honor.

Forged & Ground Brothers

The many fashions of the day that encourage countless knives flash like a Fourth of July rocket for a while, excite some for a moment, and then most fade to obscure trinkets of the past. The more significant events in the world of knives move much more slowly. Style and trends that last depend on more than the temporary nature of current interest. The lever that is the longest is the most powerful but moves with the least speed. Considering that the forged blade has been around for some time, I feel that it's safe to predict that she's here to stay. Her companions, the stainless-steel and stock-removal blades, have also firmly established themselves and will be right along with their forged sisters in the future.

There's a place and a use for every knife. In preparing to write this article, I quartered an apple with my beloved carbon-steel Frank Richtig culinary duty knife. The blade has cut my steak and done most of the kitchen stuff I have to do for more than 20 years. Her companion is a paring knife proudly stamped "Forgecraft® High Carbon" given to me by my grandmother in 1958. Each time I use the knives, I carefully wash, dry, and replace them in their wall-hanging rack. Average time required for care is about five seconds.

Angie, on the other hand, uses a stainless-steel plastic-handle knife made by Ginsu for many of her kitchen chores. As these words are written, she's gone to bed and won't return to the kitchen for several hours. Her knife is soaking in a pan filled with soapy water, as it has on many occasions, without any damage to the blade or handle. We have and closely follow an unwritten rule around the Willow Bow: Angie doesn't use my Richtig knife, and I don't fool with her stainless blade. Is one knife better than the other? I think not. Each serves a purpose, and her love for her stainless knife has a definite, positive influence on the longevity of my carbon-steel knives. (For serious work, Angie also uses kitchen knives made by Warther and Son of Dover, Ohio, as well as a Murray Carter masterpiece, and takes very good care of them.)

> "All have their place firmly established in the world of knives, and there's plenty of room for all."

Methods Of Manufacture

Along with the stainless vs. forged-carbon-steel-blade debate, there also exists a second issue: the methods of manufacture. The stock-removal makers have the option of using either carbon or stainless steel. There are also those who forge stainless steel. Many debates have taken place and many are to follow as to which method or material produces the best knife. Most debates speak in general terms, such as the age-old debates and wars over good vs. evil, the good points of steel "A" vs. the bad qualities of steel "B," as if all knives from "A" or "B" are the exact same quality simply due to their nature. These debates fail to take into consideration the most significant contributing feature to quality: the knifemaker.

Many premier makers nurture their dreams of excellence and achieve great personal satisfaction using their method of choice. They can be found working in the residential apartments of New York City to the jungles of the Philippines, on many different frontiers, all doing what they want to do—making knives. All their blades are of a different

Many debates have taken place and many are to follow as to which method or material produces the best knife. These debates fail to take into consideration the most significant contributing feature to quality: the knifemaker.

nature but each serves a purpose, be it entry-level pieces sold at carnivals or high-ticket knives displayed in satin-lined showcases.

There's nothing to be lost and much to be gained by the maker simply stating the true nature of his knives. Some will flex, others won't. Some will hold an edge through extreme use, others won't. Some are easy to sharpen, some aren't. Some will rust, others won't—or at least they will "stain less." The client who knows the performance limitations of his purchase makes it with the knowledge of what he can expect. This is called truth, and truth builds confidence, friends, and repeat customers.

Meanwhile, knifemakers can and probably should debate issues concerning which knife is best for what purpose and at what price, and continually seek to improve on their methods by sharing information freely. This is healthy for their personal growth, as well as helping to provide better knives for the future. Many dedicated men make pieces of carbon, stainless, and forged steel. All have their place firmly established in the world of knives and there's plenty of room for all. The rules are simple: Knifemakers must be as honest as they can, always remembering that in the exact geographical center of the world of knives is man, the object of all their efforts. Moreover, knifemakers must explore what frontiers they find interesting and, last but not least, enjoy their voyage in the world of knives.

Knifemakers are done a great disservice when any knifemaking method or steel is categorically criticized. The arguments tend to take the form of the age-old battle of good vs. evil. Nothing will be solved in debates of this nature and the cost is great, for all will be tarnished in the battle. There's plenty of room for any edged tool that comes from the heart of a knifemaker who seeks to do his best, be it his first blade or one of many. The terrorist in the world of sharp lies not in materials or methods, but in the misrepresented knife.

> "The terrorist in the world of sharp lies not in materials or methods, but in the misrepresented knife."

Chapter 3

Forging, Grinding and Heat Treating

This is the technical section, written in such a manner that all who read it will gain an understanding of the foundation of the High Performance Knife. This is a place where the explorer can travel many frontiers that will be new to him, discovering on his own many facets of the forged blade that may have been explored before but quietly await his visit.

Every piece of steel has a potential level of performance inherent in its chemical and physical makeup. Many events influence this potential. In the manufacturing process, cleanliness of equipment, management of the thermal cycles during the mixing of the elements, and the initial forming of the steel all play a part in the nature of the finished blade. The bladesmith has little control of these events, but through the careful selection, he can greatly reduce the influence of low quality steel by the process of elimination. The assistance of a metallurgical laboratory can be of great benefit to the bladesmith, but unless the metallurgist is familiar with the qualities of the forged blade, his contribution may be limited. The bladesmith has all the equipment necessary to experimentally examine the quality of the finished blade in his shop. The simple tests of slicing hemp rope (cut), the edge flex, and the 90-degree flex (strength and toughness), accompanied by his curiosity will take the bladesmith a long way to understanding the qualities of his blade.

Once the high quality steel is identified and selected, as long as the bladesmith does nothing to detract from the potential performance qualities of his steel and through careful planning develops the potential to its maximum level of performance, he has done all he can to achieve excellence.

"It's the beginning, not the end, that puts the edge on a knife," goes the Ballad of Charles Goodnight by Andy Wilkerson. Truer words were never spoken. Once the right steel has been selected, the nature of the High Performance Knife begins with the first heat, the first hammer blow, and continues to build to the final stroke on the sharpening stone. Like grains of wheat filling a bin, each one by itself contributes only its portion to the whole. When the many millions of grains come together to be a total, they can weigh tons. This analogy best describes the nature of the High Performance Knife: many events done right, none taking away from the quality of the finished blade and all contributing their own special quality, each infinite contribution brings the bladesmith closer in his quest for Excalibur.

We had to fight a lot of tradition to achieve our goals thus far, not necessarily the traditions of the past, but those of today. There are no secrets revealed in this section. They have been well known for centuries in the shops of individual bladesmiths working on their own, each discovering for himself the techniques that lead to their personal Excalibur.

There was a time when tools were judged not by judges on the basis of cosmetics but by men using the knives for the purpose they were made and chosen to achieve. The maker of the knife was judged solely on the high performance qualities of his product. Those who achieved the high performance blade were immediately recognized and rewarded, for their product was in demand. This is the legacy some of us choose to seek.

We have no excuse for mediocre performance. The bladesmith of today has at his call the finest steel ever available. Ours is the best of times. There is no end to the pursuit of Excalibur. The more you come to know the steel, the higher the mark, always within sight but just above your reach. This is the joy of knife making.

Not everyone agrees with me, and this is a good thing, for debate completes ideas. Included are two point-to-point discussions that were a part of the saga of the forged blade.